Trichoderma: *Ganoderma* Disease Control in Oil Palm

A Manual

Techniques in Plantation Science Series

Series editors:

Brian P. Forster, Lead Scientist, Verdant Bioscience, Indonesia
Peter D.S. Caligari, Science Strategy Executive Director, Verdant Bioscience, Indonesia

About the series:

A series of manuals covering techniques in plantation science that form the essential underlying needs to carry out plantation science.

The series reflects the expertise in Verdant Bioscience that underlies the plantation science activities carried out at the Verdant Plantation Science Centre at Timbang Deli, Deli Serdang, North Sumatra, Indonesia.

Titles available:

1. **Crossing in Oil Palm: A Manual** – Umi Setiawati, Baihaqi Sitepu, Fazrin Nur, Brian P. Forster and Sylvester Dery
2. **Seed Production in Oil Palm: A Manual** – Eddy S. Kelanaputra, Stephen P.C. Nelson, Umi Setiawati, Baihaqi Sitepu, Fazrin Nur, Brian P. Forster and Abdul R. Purba
3. **Nursery Screening for *Ganoderma* Response in Oil Palm Seedlings: A Manual** – Miranti Rahmaningsih, Ike Virdiana, Syamsul Bahri, Yassier Anwar, Brian P. Forster and Frédéric Breton
4. **Mutation Breeding in Oil Palm: A Manual** – Fazrin Nur, Brian P. Forster, Samuel A. Osei, Samuel Amiteye, Jennifer Ciomas, Soeranto Hoeman and Ljupcho Jankuloski
5. **Bunch and Oil Analysis of Oil Palm: A Manual** – Pujo Widodo, Fazrin Nur, Evi Nafisah, Brian P. Forster and Hasrul Abdi Hasibuan
6. **Nursery Practices in Oil Palm: A Manual** – Nur Dian Laksono, Umi Setiawati, Fazrin Nur, Miranti Rahmaningsih, Yassier Anwar, Heru Rusfiandi, Eben Haeser Sembiring, Brian P. Forster, Avasarala Sreenivasa Subbarao and Hafni Zahara
7. ***Trichoderma: Ganoderma* Disease Control in Oil Palm: A Manual** - Ike Virdiana, Miranti Rahmaningsih, Brian P. Forster, Monika Schmoll and Julie Flood

Trichoderma: *Ganoderma* Disease Control in Oil Palm

A Manual

Ike Virdiana,
Verdant Bioscience, Indonesia

Miranti Rahmaningsih
Verdant Bioscience, Indonesia

Brian P. Forster
Verdant Bioscience, Indonesia

Monika Schmoll
Austrian Institute of Technology GMBH, Austria

Julie Flood
CAB International, UK

CABI is a trading name of CAB International

CABI	CABI
Nosworthy Way	745 Atlantic Avenue
Wallingford	8th Floor
Oxfordshire OX10 8DE	Boston, MA 02111
UK	USA

Tel: +44 (0)1491 832111
Fax: +44 (0)1491 833508
E-mail: info@cabi.org
Website: www.cabi.org

Tel: +1 (617)682-9015
E-mail: cabi-nao@cabi.org

A catalogue record for this book is available from the British Library, London, UK.

ISBN-13: 9781789241457 (paperback)
9781789241464 (ePDF)
9781789241471 (ePub)

Commissioning Editor: Rebecca Stubbs
Editorial Assistant: Emma McCann
Production Editor: James Bishop

Typeset by SPi, Pondicherry, India
Printed and bound in the UK by Severn, Gloucester

Series Foreword – Techniques in Plantation Science

Verdant Bioscience, Singapore (VBS), is a new company established in October 2013 with a vision to develop high-yielding, high-quality planting material in oil palm and rubber through the application of sound practices based on scientific innovation in plant breeding. The approach is to fuse traditional breeding strategies with the latest methods in biotechnology. These techniques are integrated with expertise and the application of sustainable aspects of agronomy and crop protection, alongside information and imaging technology which not only find relevance in direct aspects of plantation practice but also in selection within the breeding programme. When high-yielding planting material is allied with efficient plantation practices, it leads to what may be termed 'intensive sustainable' production. At the same time, the quality of new products is refined to give more specialized uses alongside more commodity-based oil production, thus meeting the market demands of the modern world community, but with a minimal harmful footprint. An essential ingredient in all this is having sound and practical protocols and techniques to allow the realization of the strategies that are envisaged.

To achieve its aims, VBS acquired an Indonesian company called PT Timbang Deli Indonesia, with an estate of over 970 ha of land at Timbang Deli, Deli Serdang, North Sumatra, Indonesia, and the group works under the name of 'Verdant'. A central part of this estate, which will be used for important plant nurseries and field trials, is the development of the Verdant Plantation Science Centre (VPSC), to which the operational staff moved in October 2016. A seed production and marketing facility is now established at VPSC for commercial seed sales and the processing of seed from breeding programmes. The centre comprises specialized laboratories in cell biology, genomics, tissue culture, pollen, soil DNA, plant and soil nutrition, bunch and oil, agronomy and crop protection. Field facilities include extensive nurseries, seed gardens and trials (trial sites are also located at various locations across Indonesia). It is the aim of the company to use its existing and rapidly developing intellectual property (IP) to develop superior cultivars that not only have outstanding yield but also are resistant to both biotic and abiotic stresses, while at the same time meeting new market demands. Verdant not only develops and supplies superior planting materials but also supports its customers and growers with a package of services and advice in fertilizer recommendations and crop protection. This is all part of a central mission to promote green, eco-friendly agriculture.

Brian P. Forster and Peter D.S. Caligari
Lead Scientist and Science Strategy Executive Director
Verdant Bioscience

Contents

Preface to manual: *Trichoderma*: Disease Control in Oil Palm

As noted in the Foreword to this Series, a central objective in Verdant Bioscience's mission is to breed and sell better, more sustainable varieties of oil palm, rubber and other plantation crops, through plant breeding. An essential part of introducing more sustainable varieties is the package of advice and techniques relevant to the new planting material and its subsequent development in the plantation. Thus, one of Verdant's services is to provide advice, recommendations and training to growers. Basal stem rot (BSR) caused by *Ganoderma boninense* is the major disease of oil palm in S.E. Asia. The occurrence of this disease is increasing and is particularly problematic in relation to re-planting oil palm with further rounds of oil palm. The implementation of a fallow period is being increasingly accepted, despite the initial economic impact, as a major sustainable practice, particularly as the disease is soil-borne. BSR can cause up to 80% of palm losses in a plantation especially in later rounds of re-planting. Currently there is little variation in terms of resistance or tolerance to the disease in commercial material, although there are some companies who provide material claiming a modest level of tolerance. So, realistic disease control must rely on agronomic practices, such as fallow periods, but also more recently introduced is the soil application of *Trichoderma* during field planting. *Trichoderma* is a soil fungus that is antagonistic to *Ganoderma*. This manual describes methods for isolating *Trichoderma* from the wild, *in vitro* culture methods, assessing virulence in combatting *Ganoderma*, commercial production and application to the soil during field planting of nursery-produced plants. The manual is one in a series of manuals on Techniques in Plantation Science and has direct relevant links with other manuals in the series, particularly: *Nursery Practices in Oil Palm*, *Field Trialling in Oil Palm* and *Nursery Screening for* Ganoderma *Response in Oil Palm Seedlings*. Our target audiences are students and researchers in agriculture, plant pathology, agronomy, growers and end-users interested in the practicalities of sustainable oil palm production. It provides a resource for training, a knowledge base for those new to oil palm and a reference guide to managers, particularly if they are interested in increasing the sustainability of their production. Our aim is to help in providing best practices in maximising sustainability and production of this important and valuable crop.

Brian P. Forster and Peter D.S. Caligari

Acknowledgements

The authors are grateful to colleagues in Verdant Bioscience Breeding and Crop Protection Departments for sharing their knowledge and providing helpful advice in preparing this manual.

Introduction

1

Abstract

Trichoderma species are filamentous fungi found colonizing various habitats including plant materials such as stems, leaves, fruits and roots. *Trichoderma* species are typically found in soil associated with plant roots. The genus is of considerable interest for their its potential in the management of plant diseases. For about 70 years, *Trichoderma* spp. have been known to attack other fungi, to produce antibiotics that affect other microbes and to act as bio-control agents. Biological control using *Trichoderma* is part of an integrated approach to manage 'Basal Stem Rot' (BSR) disease of oil palm. BSR, caused by *Ganoderma* species, is considered to be the greatest threat to oil palm production in Southeast Asia (the largest area of production of palm oil). The use of *Trichoderma* is part of integrated disease management systems that have been developed to reduce the impact of BSR disease. This management approach consists of a combination of husbandry and agronomic techniques, chemical, and biological control (especially *Trichoderma* application to the soil prior to field planting) and to develop resistant material through breeding and biotechnological approaches. Currently there is little resistance to *Ganoderma* in breeding materials and it will take time to develop such resistant materials, in which probably the resistance will never be complete but rather will be partial, so the use of *Trichoderma* as a biological control is therefore critical in the management of the disease both in the shorter-term period as well as being a longer-term part of integrated disease management systems. This introductory chapter describes the biology of *Trichoderma*, its ability to attack other fungi and how this can be harnessed as a biological control measure for BSR.

1.1   *Trichoderma* – Biology, Life-cycle and Antagonism

Species of the genus *Trichoderma* belong to one of the most useful groups of microbes and have many applications (Mukherjee *et al.*, 2013). They

have been widely used as bio-fungicides and plant growth modi-
fiers as well as being sources of enzymes for industrial use. In soil,
Trichoderma species are used in the bioremediation of organic and in-
organic wastes including heavy metals (Schuster and Schmoll, 2010;
Harman, 2011a, 2011b). They are able to colonize plant parts including stems, leaves, fruits and roots; some species have been shown
to grow endophytically (Bailey and Melnick, 2013). *Trichoderma* species prefer locations with a large supply of plant roots, which they
promptly colonize. In addition, *Trichoderma* species attack, parasitize
or derive nutrition from other fungi (Reetha *et al.*, 2014). Although
Trichoderma species have been considered as soil inhabitants, based on
in situ diversity studies using a taxon-specific metagenomics approach,
Friedl and Druzhinina (2012) suggest that only a relatively small number
are adapted to soil as a habitat. Properly selected isolates interact with
the plant by colonizing roots, establishing chemical communication and
systemically altering the expression of numerous plant genes.

Trichoderma is easily cultured, although a specific pH is required for
maximum growth where these biocontrol agents can be multiplied and target
pathogens controlled. Studies of pH variation by different workers revealed
that *Trichoderma* isolates show optimum growth and sporulation rates at
different pH values ranging from 2 to 7 (Begoude, 2007). Earlier experiments in India showed that the most favourable pH ranges were between
5.5 and 7.5 in which total dry weight of mycelium varies between 1.41 and
1.35 g. Although all the species of *Trichoderma* produced sufficient biomass
at different temperatures viz. 20°C, 25°C, 30°C and 35°C, they were found
to grow best at a temperature range of 25°C to 30°C. Aeration by agitation
was also studied, with the greatest biomass recorded at 150 rpm (Singh
et al., 2014).

With respect to life cycle, for many years *Trichoderma* isolates were
considered asexual (anamorph) clonal lines of formally sexually repro-
ducing species (teleomorphs). However, it was suggested by Tulasne and
Tulasne (1865, see Schuster and Schmoll, 2010) that there was a link to
the sexual state of *Hypocrea* species, this link was proven experimentally
by Seidl *et al.*, (2009). The limited observations of the sexual stage and the
reasons for this are discussed in more detail by Schmoll (2013). The or-
ganism grows and ramifies as typical fungal hyphae, 5 to 10 µm in diameter.
Asexual sporulation occurs as single cells, usually green conidia (typically
3 to 5 µm in diameter) that are released in large numbers on a repetitively
branched conidiophore structure. Co-intercalary resting chlamydospores
are also formed, these are also single celled, although two or more chlamydospores may be fused together (Gams and Bissett, 1998; Singh *et al.*,
2012).

Relatively recently, *Trichoderma* isolates have been identified as being
able to act as endophytic plant symbionts. The isolates become endophytic
in roots, but the greatest changes in gene expression occur in shoots. These

changes alter plant physiology and may result in the improvement of abiotic stress resistance, nitrogen fertilizer uptake, resistance to pathogens and photosynthetic efficiency. Typically, the net result of these effects is an increase in plant growth and productivity (Hermosa *et al.*, 2012). They are also able to induce disease suppression in soils, and this is currently their major use in oil palm. The complex mechanisms of mycoparasitism, which include direct growth of *Trichoderma* toward target fungi, attachment and coiling of *Trichoderma* on target fungi, and the production of a range of antifungal extracellular enzymes have been described (Chet, 1987; Chet *et al.*, 1998; Zeilinger and Omann, 2007).

1.2   *Trichoderma* as a Biological Control Agent

Trichoderma spp. produce a wide range of extracellular enzymes, some of which have been implicated in the biological control of plant diseases (Elad *et al.*, 1982; Zeilinger and Omann, 2007). The enzymes themselves were found to be toxic to fungi and mixtures of enzymes were synergistic in their antifungal properties. Different classes of chitinolytic or glucanolytic enzymes from *Trichoderma* are synergistic, as are enzymes from different organisms (Lorito *et al.*, 1996).

Efficient bio-control isolates of the genus are being developed as promising biological fungicides, and their weaponry for this function also includes secondary metabolites with potential applications as novel antibiotics (Schuster and Schmoll, 2010). They are able to deal with such different environments as the rich and diversified habitat of a tropical rain forest as well as with the dark and sterile setting of a biotechnological fermentor or shaker flask. Under all these conditions, they respond to their environment by regulation of growth, conidiation, enzyme production, and hence adjust their lifestyle to current conditions, which can be exploited for the benefit of mankind. One of these environmental factors is the presence or absence of light (Schmoll *et al.*, 2010).

Trichoderma spp. have also been characterized as opportunistic avirulent symbionts (Harman *et al.*, 2004). The critical characteristic of this association is the penetration of the plant's root system by *Trichoderma* and the persistent survival of the fungus within living plant tissues. Recent research results, principally with cocoa, *Theobroma cacao*, demonstrate that *Trichoderma* species can persist not only within the plant's root system but also within above-ground tissues in endophytic associations (Evans *et al.*, 2003; Bailey *et al.*, 2006; Bailey *et al.*, 2008).

Trichoderma species can be re-isolated from surface sterilized cocoa stem tissue, including the bark and xylem, the apical meristem, and to a lesser degree from leaves. Scanning Electron Microscopy (SEM) analysis of cocoa stems colonized by isolates of four *Trichoderma* species (*Trichoderma ovalisporum*-DIS 70a, *Trichoderma hamatum*-DIS 219b, *Trichoderma koningiopsis*-DIS 172ai, or *Trichoderma harzianum*-DIS 219f) showed a preference for surface colonization of glandular trichomes versus non-glandular trichomes. The *Trichoderma* isolates colonized the glandular trichome tips and formed swellings resembling appressoria. Hyphae were observed emerging from the glandular trichomes on surface sterilized stems from cocoa seedlings that had been inoculated with each of the four *Trichoderma* isolates. Fungal hyphae were observed under the microscope emerging from the trichomes as soon as 6 h after their isolation from surface sterilized cocoa seedling stems. Hyphae were also observed, in some cases, emerging from stalk cells opposite the trichome head. Repeated single trichome/hyphae isolations verified that the emerging hyphae were the *Trichoderma* isolates with which, the cocoa seedlings had been previously inoculated. Isolates of four *Trichoderma* species were able to enter glandular trichomes during the colonization of cocoa stems where they survived surface sterilization and could be re-isolated. The penetration of cocoa trichomes may provide the entry point for *Trichoderma* species into the cocoa stem allowing systemic colonization of this tissue (Bailey *et al.*, 2009).

Many of the endophytic *Trichoderma* species isolated from cocoa environments are being studied for their potential to protect cocoa against diseases. Black Pod (*Phytophthora* species), Witches' Broom (*Moniliophthora perniciosa*) and Frosty Pod Rot (*Moniliophthora roreri*) are major cocoa diseases that colonize above-ground tissues. All three diseases occur in South and Central America although their distributions vary, as does their relative importance (Bowers *et al.*, 2001; Wood and Lass, 2001).

1.3   Basal Stem Rot Disease in Oil Palm

Basal stem rot (BSR) was first reported in the Congo (now Democratic Republic of the Congo) in 1915 (Wakefield, 1920), and although the disease kills infected palms the incidence was considered too low to be a serious threat to crop production (Turner, 1981). In Southeast Asia, BSR caused by various species of *Ganoderma* is the only disease that is causing serious losses in field plantings, especially in Malaysia and Indonesia (Susanto, 2009a). Although the disease has been recorded in Africa, Central America and Papua New Guinea, its impact is less significant (Turner and Gillbanks, 2003). Up to 80% yield losses have been recorded with an estimated loss in revenue of US$ 500 million per year (Breton *et al.*, 2010; Hushiarian *et al.*, 2013). The disease is a problem on both mineral and peat soils.

In Indonesia, especially North Sumatra, where most of oil palm plantations are already in their second or third replants, *Ganoderma* incidence has increased dramatically. Furthermore, zero burning, as noted in RSPO (Roundtable on Sustainable Palm Oil) regulation, during the first generation of planting (from forest) limited the reduction of disease sources on the old forest trees which are infected by root disease, especially *Ganoderma*. In addition, there was only limited understanding of the disease and its destructive potential for oil palm; now we are more aware. In the first generation, some *Ganoderma* incidence was observed but it was considered to be only a minor problem and restricted to older oil palms. However, at replanting (at end of first generation) most *Ganoderma* infection is in the bottom part of the oil palm stem and in the bole, therefore the use of a windrowing system of this infected material provided inoculum for subsequent generations. Disease symptoms were seen earlier in the planting with the pathogen not only infecting older palms, but also young palms.

Chung (2011) estimated the economic loss of a single palm due to basal stem rot at ages of 10, 15 and 20 years to be RM 2100 ($508), RM 1400 ($339) and RM 700 ($169) respectively, so infection with the pathogen at an earlier stage in the planting has significant economic impact. In one oil palm estate in North Sumatra, Indonesia, a census was conducted in 2011, using Global Positioning System (GPS) in order to record the percentage *Ganoderma* infection of palms <6 years old, as 0-1.5%. Percentage infection increased dramatically in palms >16 years old to 13-87% (this included vacant points which usually would have been as a result of *Ganoderma* infection) with a stand per hectare ranging from 35 to 119 palms in the worst affected fields (Virdiana *et al.*, 2012b).

Consequently, BSR is particularly devastating in plantations where oil palm is grown continuously. Successive re-planting, a common practice, brings about early and more frequent disease incidence with a major impact on yield (Breton *et al.*, 2010; Purba *et al.*, 2012). Indonesia and Malaysia are the two main producers of palm oil, and the main countries impacted by *Ganoderma* infection, although infection is increasing now in PNG (Papua New Guinea) too (Pilotti *et al.*, 2004). There is therefore a concerted effort in seeking methods to manage this devastating disease (Hama-Ali *et al.*, 2014; Rakib *et al.*, 2015).

1.4   Causes of Basal Stem Rot Disease

Root and stem rots caused by *Ganoderma boninense* were first described in 1915 in the Congo, West Africa, as a disease of senescing palms (Wakefield, 1920). The disease was first identified in Malaysia by Thomson (1931). In many parts of the world, fifteen species of *Ganoderma* have been recorded as likely pathogens (Turner, 1981). Ho and Nawawi (1985) concluded that all *Ganoderma* isolates from diseased oil palms from various locations in Malaysia were all the same species, *G. boninense*. However, morphological characters of the basidiomata suggest other species of *Ganoderma* were involved, namely *G. boninense, G. miniatocintum* and *G. zonatum* (Khairudin, 1990; Idris and Ariffin, 2004).

In Indonesia, the oil palm industry has developed rapidly since 2011, replacing Malaysia as the world's biggest producer in the world. Areas of oil palm expansion in Indonesia include Sumatra, Kalimantan, Sulawesi, Papua and West Java. The expansion was sometimes opening forest areas, but also included large areas that were conversions from other plantation crops (Lubis, 2009).

Fig. 1.1. *Ganoderma* disease symptoms: (a) Early infection (b) Advanced infection (c) Fallen palm (d) *Ganoderma* basidiocarp

1.5   Control of Basal Stem Rot Caused by *Ganoderma*

Although finding the best solution of BSR disease of oil palm is not easy, some efforts have been developed and practised to reduce infection of *Ganoderma* disease. No single method can claim to give a hundred percent success, and therefore, integrated disease management is recommended. This comprises chemical, biological and agronomic controls. Breeding efforts are also ongoing to produce resistant oil palm, but so far have only been partially successful.

(a) Cultural Control

Mounding of soil around infected palms has been utilized as a method of extending the productive life of infected palms. This stimulates renewed root development from stem tissue above the infected region and provides added stability, which compensates for reduced stem structural integrity due to decay caused by *G. boninense*. However, this treatment simply extends the life of the palm and does nothing to prevent the course of infection or transmission of the fungus (Tuck and Hashim, 1997) and could actually increase the inoculum.

Fig. 1.2. Mounding at the base of a lightly infected *Ganoderma* palm

At re-planting, shredding the oil palm stems (into slices about 10 cm thick) significantly reduced *Ganoderma* infection in the oil palm re-plant (Virdiana, 2012a). This encourages oil palm material to decay more quickly and any *Ganoderma* present in the debris becomes vulnerable to antagonism by natural microbial populations in the soil. *Ganoderma* is not a good competitor in soil (Rees, 2006).

A one-year fallow also significantly reduces *Ganoderma* infection as it reduces inoculum, again due to microbial antagonism (competition) and lack of a suitable host. However, a fallow period before re-planting will have a major impact on Internal Rates of Return (IRRs) and on an oil palm plantation's cash flow. It is therefore necessary to conduct further work to assess the optimal and minimal fallow periods to determine the best re-planting

Fig. 1.3. Shredding oil palm stems (into slices about 10 cm thick)

time. Commercially, a viable cash crop (e.g. banana, maize, legumes) could be planted as a "break crop" (Virdiana, 2012a). Soils may be sampled for DNA extraction to monitor the incidence of *Ganoderma* and other microbes in the soil, so determining soil health.

(b) Chemical control

Although many fungicides have been shown to be effective in supressing *Ganoderma* in laboratory experiments, their effectiveness in the field is poor (Susanto, 2002; Hushiarian *et al.*, 2013). Rees (2006) stated there is inconclusive evidence for efficacy in reducing disease incidence or in prolonging the productive life of infected trees. However, application of the systemic fungicide triadimenol was claimed to prolong the life of infected palms (Ariffin *et al.*, 2000). Consequently, chemical control of BSR has been limited and more research is required to improve this approach.

(c) Breeding

Breeding programmes for *Ganoderma* resistance have been set up in Indonesia and Malaysia. Success would be a major advance in protecting oil palm from this disease (Susanto, 2002; Breton *et al.*, 2010; Purba *et al.*, 2012). However, there are few potent sources for disease resistance, and little is known about the genetics of resistance. Some Indonesian seed producers have released *Ganoderma*, tolerant planting material, but this is a low level of partial tolerance and IPM (Integrated Pest Management) and other additional management methods are required.

(d) Biological Control

Trichoderma has been used as a biological control of *Ganoderma* since at least the 1920s (Harman, 2006). For oil palm, *Trichoderma* is multiplied and produced on a large scale, then used as a soil treatment. The application methods such as drenching of the new plant and manual application into the planting hole during replanting can be done as a prevention method for the root disease problem. Screening for the most effective isolates of *Trichoderma* may be conducted effectively in the nursery.

Nursery seedling trials (in pots) have shown that the highest *Ganoderma* infection (up to 100%) occurred in treatments with no *Trichoderma*. Infection was significantly higher when compared to treatments where *Trichoderma* was applied on top of rubber wood blocks (*Ganoderma* inoculum source in the pot) while the lowest infection was in treatments with *T. koningii*, *T. harzianum* and *T. virens* (Virdiana *et al.*, 2012b). In addition, *Trichoderma* effectiveness has not only been demonstrated in the laboratory and in the nursery, but also significantly under field conditions. *Trichoderma* application to the soil is now implemented by a growing number of plantation companies. For example, Priwiratama and Susanto (2014) reported that the application of *T. virens* to planting holes is conducted routinely in many oil palm re-planting schemes.

The general methods involved are illustrated in Figure 1.4 and described in detail in Chapters 9 and 10.

Fig. 1.4. *Trichoderma* applications: (a) mixing soil with *Trichoderma* in the nursery (b) Manual application in the planting hole

1.6   *Trichoderma* as a Biological Control Agent for *Ganoderma*

Trichoderma isolates used in the biological control of *Ganoderma* are normally initially selected from *in vitro* studies (see Chapter 6), and from nursery testing (see Chapter 7). The ability of *Trichoderma* isolates to inhibit *Ganoderma* infection in oil palm has been observed using a glass chamber. *Trichoderma* starts to colonize *Ganoderma* isolates on rubber wood blocks

(RWB) 2 weeks after inoculation, resulting in no *Ganoderma* growth 4 weeks post inoculation with the more aggressive *Trichoderma* strains. Seedlings infected by *Ganoderma* were observed in the control treatment (no *Trichoderma*) 3 months post inoculation while those seedlings treated with *Trichoderma* showed no signs of *Ganoderma* infection 24 months post inoculation (Anjara *et al.*, 2011; Anjara *et al.*, 2013).

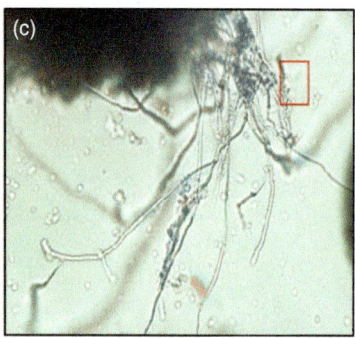

Fig. 1.5. (a) the seedling infected by *Ganoderma* in control treatment 12 weeks post inoculation; (b) seedling primary root in contact with *Ganoderma* RWB (Rubber Wood Block) as source of inoculum; (c) the photograph of *Ganoderma* mycelium taken from the root of seedling infected by *Ganoderma*; (d) seedling treated with *Trichoderma* showing no *Ganoderma* infection 24 weeks post inoculation

Microscopic studies involving *T. harzianum* showed that hyphae could penetrate *G. boninense* mycelia and so form a barrier between both mycelia. Interaction between *T. viride* and *G. boninense* was rather different with no barrier zone being observed, all *T. viride* hyphae could penetrate into *G. boninense* hyphae (Susanto *et al.*, 2002).

In vitro tests have indicated that some *Trichoderma* isolates were such aggressive mycoparasites that they completely colonized *G. boninense* mycelium such that the pathogen could not be recovered - indicating the pathogen had been killed (Sundram, 2013). *Trichoderma* pre-treated seedlings (four months) were planted at four corners around *Ganoderma* infected stumps. Seedlings that were pre-treated with two kinds of *T. virens* had less disease symptoms compared to untreated positive controls where no *Trichoderma* pre-treatment had been conducted (Sundram *et al.*, 2016). In Verdant, bait seedlings are used to screen the best isolates in the field. Three or four bait seedlings are planted around a palm with a *Ganoderma* score one (light *Ganoderma*) palm combined with *Trichoderma* application into the planting hole to assess the best isolates to control *Ganoderma* in bait seedlings.

References

Anjara, P., Virdiana, I., Flood, J., Ritchie, B.J. and Nelson, S. (2011) Preliminary in vitro and nursery results to screen for *Trichoderma* isolates antagonistic to *Ganoderma*. In: Proceedings of the International Palm Oil (PIPOC) Congress, Agriculture, Biotechnology and Sustainability. 15th-18th November. Kuala Lumpur: Malaysian Palm Oil Board, pp. 424-427.

Anjara, P., Breton, F., Belson, S.P.C., Rahmaningsih, M., Setiawati, U., Virdiana, I. and Flood, J. (2013) Some approaches to *Ganoderma* Management in Sumatra. (PIPOC) Congress, Agriculture, Biotechnology and Sustainability (oral presentation). 20th-21th November. Kuala Lumpur: Malaysian Palm Oil Board.

Ariffin, D., Idris, A.S. and Singh, G. (2000) Status of *Ganoderma* in oil palm. In: Flood, J., Bridge, P. and Holderness, M. (Eds.). *Ganoderma* Disease of Perennial Crops. CAB International, Wallingford, UK, pp. 49-68.

Bailey, B.A. and Melnick, R.L. (2013) The Endophytic *Trichoderma*. *In:* Mukherjee, P.K., Singh, U.S., Horwitz, B.A. and Schmoll, M. (Eds). *Trichoderma*: biology and applications. CAB International, Wallingford, UK, pp. 152-172.

Bailey, B.A., Bae, H., Strem, M.D., Roberts, D.P., Thomas, S.E., Crozier, J., Samuels, G.J., Choi, I-Y. and Holmes, K.A. (2006) Fungal and plant gene expression during the colonization of cacao seedlings by endophytic isolates of four *Trichoderma* species. Planta 224, 1449-1464.

Bailey, B.A., Bae, H., Strem, M.D., Crozier, J., Thomas, S.E., Samuels, G.J., Vinyard, B.T. and Holmes, K.A. (2008) Antibiosis, mycoparasitism, and colonization success for endophytic *Trichoderma* isolates with biological control potential in *Theobroma cacao*. Biological Control 46, 24-35.

Bailey, B.A., Strem, M.D. and Wood, D. (2009) *Trichoderma* species form endophytic associations within *Theobroma cacao* trichomes. Mycological Research 113, 1365-1376.

Begoude, B.A., Lahlali, R., Friel, D., Tondje, P.R. and Jijakli, M.H. (2007) Response surface methodology study of the combined effects of temperature, pH, and aw on the growth rate of *Trichoderma asperellum*. Journal of Applied Microbiology 103, 845-854.

Bowers, J.H., Bailey, B.A., Hebbar, P.K., Sanogo, S. and Lumsden, R.D. (2001) The impact of plant disease on world chocolate production. Plant Health Progress. doi:10.1094/PHP-2001-0709-01-RV (online).

Breton, F., Hasan, Y., Hariadi, Lubis, Z. and De Franqueville, H. (2006) Characterization of parameters for the development of an early screening test for basal stem rot tolerance in oil palm progenies. Journal of Oil Palm Research Special Issue April 2006, 24-36.

Breton, F., Rahmaningsih, M., Lubis, Z., Syahputra, I., Setiawati, U., Flori, A., Sore, R., Jacquemard, J., Cochard, B., Nelson, S., Durand-Gasselin, T. and De Franqueville, H. (2010) Evaluation of resistance/susceptibility level of oil palm progenies to basal stem rot disease by the use of an early screening test, relation to field observations. Second International Seminar Oil Palm diseases - Advances in *Ganoderma* Research and Management, 31 May 2010 in Yogyakarta, Indonesia.

Chet, I. (1987) *Trichoderma*-application, mode of action, and potential as a biocontrol agent of soilborne plant pathogenic fungi. In: Chet, I. (Ed) Innovative approaches to plant disease control. New York: John Wiley and Sons, pp. 137-160.

Chet, I., Benhamou, N. and Haran, S. (1998) Mycoparasitism and lytic enzymes. In: Harman, G.E. and Kubicek, C.P. (Eds) *Trichoderma* and *Gliocladium*, Volume 2: Enzymes, Biological Control and commercial applications. London: Taylor and Francis, pp. 153-172.

Chung, G.F. (2011) Management of *Ganoderma* diseases in oil palm plantations. The Planter 87(1022), 325-339.

Elad, Y., Chet, I. and Henis, Y. (1982) Degradation of plant pathogenic fungi by *Trichoderma harzianum*. Canadian Journal of Microbiology 28(7), 719-725.

Evans, H.C., Holmes, K.A. and Thomas, S.E. (2003) Endophytes and mycoparasites associated with an indigenous forest tree, *Theobroma gileri*, in Ecuador and a preliminary assessment of their potential as biocontrol agents of cocoa diseases. Mycological Progress 2, 149-160.

Friedl, M.A. and Druzhinina, I.S. (2012) Taxon-specific metagenomics of *Trichoderma* reveals a narrow community of opportunistic species that regulate each other's development. Microbiology 158, 69-83.

Gams, W. and Bissett, J. (1998) Morphology and identification of *Trichoderma*. In: Harmann, G.E. and Kubicek, C.P. (Eds) *Trichoderma* and Gliocladium, Volume 1: Basic Biology, Taxonomy and Genetics. London: Taylor and Francis, pp. 3-34.

Hama-Ali, E.O., Panandam, J.M., Soon, G.T., Alwee, S.S.R.S., Tan, J.S., Ho, C.L., Namasivayam, P. and Hoh, B.P. (2014) Association between basal stem rot disease and simple sequence repeat markers in oil palm, *Elaeis guineensis* Jacq. Euphytica 202(2), 199-206.

Harman, G.E. (2006) Overview of mechanism and uses of *Trichoderma* spp. The American Phytopathological Society 96(2), 190-194.

Harman, G.E., Howell, C.R., Viterbo, A., Chet, I. and Lorito, M. (2004) *Trichoderma* species opportunistic, avirulent plant symbionts. Nature Reviews 2, 43-56.

Harman G. E. (2011a) Multifunctional fungal plant symbionts: new tools to enhance plant growth and productivity. New Phytologist 189, 647–649.

Harman G.E (2011b) *Trichoderma*—not just for biocontrol anymore. Phytoparasitica 39, 103–108.

Hermosa, R., Viterbo, A., Chet, I. and Monte, E. (2012) Plant-beneficial effects of *Trichoderma* and of its genes. Microbiology 158, 17–25.

Ho, Y.W. and Nawawi, A. (1985) *Ganoderma boninense* Pat. from basal stem rot of oil palm (*Elaeis guineensis*) in Peninsular Malaysia. Pertanika 8, 425-428.

Hushiarian, R., Yusof, N.A. and Dutse, S.W. (2013) Detection and control of *Ganoderma boninense*: strategies and perspectives. Springer Plus 2(555), 1-12.

Idris, A.S., Ismail, S., Ariffin, D. and Ahmad, H. (2004) Prolonging the productive life of *Ganoderma* infected palms with hexaconazole. MPOB TT No 214, pp. 4.

Khairudin, H. (1990) Basal stem rot of oil palm: incidence, etiology and control. Master Agriculture Science Thesis. University Pertanian Malaysia, Selangor, Malaysia. 152 pp.

Kubicek, C.P. and Harman, G.E. (1998) *Trichoderma* and *Gliocladium*, Volume 1: Basic Biology, Taxonomy and Genetics. Pp 270.

Lorito, M., Woo, S.L., Donzelli, B. and Scala, F. (1996) Synergistic, antifungal interactions of chitinolytic enzymes from fungi, bacteria and plants. In: Muzzarelli, R.A.A. (Ed) Chitin Enzymology, Proceedings of the 2nd International Symposium on Chitin Enzymology, Senigallia (Italy), 8-11 May, 1996. Madison: Atec, pp. 157-164.

Lubis, A.U. (2009) Kelapa Sawit (*Elaeis guineensis* Jacq) di Indonesia. Medan: Pusat Penelitian Kelapa Sawit Indonesia.

Merciere, M., Boulord, R., Carasco-Lacombe, C., Klopp, C., Lee, Y.P., Tan, J.S., Shahrul Rabiah, S., Zaremski, A., De Franqueville, H., Breton, F. and Lamus-Kulandaivelu, L. (2017) About *Ganoderma boninese* in oil palm plantations of Sumatra and peninsular Malaysia: ancient population expansion, extensive gene flow and large scale dispersion ability. Fungal Biology (2017), 1-12.

Mukherjee, P.K., Horwitz, B.A., Singh, U.S., Mukherjee, M. and Schmoll, M. (2013) *Trichoderma*: biology and applications. CAB International, Wallingford, UK, pp. 327.

Pilotti, C.A., Sanderson, F.R., Aitken, E.A.B. and Amstrong, W. (2004) Morphological variation and host range of two *Ganoderma* species from Papua Nuginea. Mycopathologia 158, 251-265.

Priwiratama, H. and Susanto, A. (2014) Utilization of fungi for the biological control of insect pests and *Ganoderma* disease in the Indonesian oil palm industry. Journal of Agricultural Science and Technology A 4 (2014), 103-111.

Purba, A.R., Setiawati, U., Rahmaningsih, M., Yenni, Y., Rahmadi, H.Y. and Nelson, S. (2012) Indonesia's experience of developing *Ganoderma* tolerant/resistant oil palm planting material. 2012 The International Society for Oil Palm Breeders (ISOPB) International Seminar on Breeding for Oil Palm Disease Resistance, 24th November 2012 in Bogota, Colombia.

Rahmaningsih, M., Setiawati, U., Breton, F., Sore, R., Nelson, S. and Caligary, P.D.S. (2010) General Combining Ability (GCA) estimates for *Ganoderma* partial resistance from nursery screening trials. In: Siahaan, D., Samosir, Y., Herawan, T., Rahutomo, S., Jatmika, A., Erwinsyah, Susanto, A., Sutarta, E.S., Panjaitan, F.R. and Hasibuan, H.A. (Eds) Proceedings of International Oil Palm Conference 2010, Agriculture, 1-3 June 2010, Yogyakarta, Indonesia. Medan: IOPRI.

Rakib, M.R.M., Bong, C.F.J., Khairulmazmi, A. and Idris, A.S. (2015) Aggressiveness of *Ganoderma boninense* and *G. zonatum* isolated from upper and basal stem rot of oil palm (*Elaeis guineensis*) in Malaysia. Journal of Oil Palm Research 27(3), 229-240.

Rees, R.W. (2006) *Ganoderma* stem rot oil palm (*Elaeis guineensis*): Mode of Infection, Epidemiology and Biological Control. PhD Thesis. University of Bath, Bath UK.

Reetha, S., Bhuvaneswari, G., Selvakumar, G., Thamizhiniyan, P. and Pathmavanthi, M. (2014) Effect of temperature and pH on growth of fungi *Trichoderma harzianum*. Journal Chemical, Biology and Physical Sciences 4(4), 3287-3292.

Schmoll, M. (2013) Sexual Development in *Trichoderma* – Scrutinizing the Aspired Phenomenon. In: Mukherjee, P.K., Singh, U.S., Horwitz, B.A. and Schmoll, M. (Eds) *Trichoderma*: biology and applications. CAB International, Wallingford, pp. 67–86.

Schmoll, M., Esquivel-Naranjo, E.U., Herrera-Estrella, A. (2010) *Trichoderma* in the light of day-physiology and development. Fungal Genetics and Biology 47, 909–916.

Seidl, V., Seibel, C., Kubicek, C.P., Schmoll M. (2009) Sexual development in the industrial workhorse *Trichoderma* reesei. Proc Natl Acad Sci USA 106:13909–13914.

Schuster, A. and Schmoll, M. (2010) Biology and biotechnology of *Trichoderma*. Applied Microbiology and Biotechnology 87(3), 787–799.

Singh, A., Shahid, M., Srivastva, M. and Kumar, V. (2012) Production of biocontrol agent *Trichoderma* in organic agriculture. Advances in Life Sciences 1(2), 100-103.

Singh, A., Shahid, M., Srivastava, M., Pandey, S., Sharma, A. and Kumar, V. (2014) Optimal physical parameters for growth of *Trichoderma* species at varying pH, temperature and agitation. Virology and Mycology 3(1), 1-7.

Singh, G. (1991) *Ganoderma* – the scourge of oil palm in the coastal areas. The Planters 67(786), 421-444.

Sundram, S. (2013) First report isolation of endophytic *Trichoderma* from oil palm (*Elaeis guineensis* Jacq.) and their *in vitro* antagonistic assessment on *Ganoderma boninense*. Journal of Oil Palm of Oil Palm Research 25(3), 368-372.

Sundram, S., Angel, L.P.L., Ping, B.T.Y., Roslan, N.D., Inam, A. and Idris, A.S. (2016) *Trichoderma virens*, an effective biocontrol agent against *Ganoderma boninense*. MPOB Information Series MPOB TT no. 587.

Susanto, A. (2002) Biological Control of *Ganoderma boninense* Pat., the Causal Agent of Basal Stem Rot Disease of Oil Palm. Doctoral Dissertation. Bogor Agricultural University, Bogor.

Susanto, A. (2009) Basal stem rot in Indonesia – biology, economic importance, epidemiology, detection and control. In: A Kushairi, AS Idris, K Norman Proceedings of International Workshop on Awareness, Detection and Control of Oil Palm Devastating Diseases, 6 November 2009, Kuala Lumpur, Malaysia (pp. 58-89). Kuala Lumpur: MPOB.

Thomson, A. (1931) Stem-rot of the oil palm in Malaya. Bulletin. Department of Agriculture, Straits Settlements and F.M.S., Science Series 6.

Tuck, H.C. and Hashim, K. (1997) Usefulness of soil mounding treatment in prolonging productivity of prime-aged *Ganoderma* infected palms. The Planter 73:239-244.

Turner, P.D. (1981) Oil Palm Diseases and Disorders. Oxford, New York, Melbourne: Oxford University Press.

Turner P.D., Gillbanks R.A. (2003) Oil palm cultivation and management, The Incorporated Society of Planters, Kuala Lumpur, Malaysia

Utomo, C. (2002) Studies on Molecular Diagnosis for Detection, Identification and Differentiation of Ganoderma, the Causal Agent of Basal Stem Rot Disease in Oil Palm. Doctoral Dissertation of Agricultural Sciences of the Faculty of Agricultural Sciences. Martin Luther University Halle – Wittenberg Germany.

Virdiana, I., Flood, J., Sitepu, B., Hasan, Y., Aditya, R. and Nelson, S. (2011) Integrated disease management to reduce future *Ganoderma* infection during oil palm replanting. Proceedings of the PIPOC 2011 International Oil Palm Congress, Kuala Lumpur, Malaysia. 130-134.

Virdiana, I., Flood, J., Sitepu, B., Hasan, Y., Aditya, R. and Nelson, S. (2012a) Integrated disease management to reduce future *Ganoderma* infection during oil palm replanting. The Planter 88(1305), 383–393.

Virdiana, I., Anjara, P., Flood, J., Sitepu, B., Hasan, Y., Aditya, R. and Nelson, S. (2012b) Replanting system and *Trichoderma* as an effective form of *Ganoderma* control and proven results. 4th Oil Palm Summit. 09-10 July. Denpasar. P1-10.

Wakefield, E.M. (1920) Disease of the Oil Palm in West Africa. Kew Bulletin: 306-308 In: Rees, RW. *Ganoderma* stem rot oil palm (*Elaeis guineensis*): Mode of Infection, Epidemiology and Biological Control. PhD Thesis. University of Bath, Bath UK, pp 244.

Weindling, R. and Fawcett, H.S. (1936) Experiments in the control of *Rhizoctonia* damping-off of citrus seedlings. Hilgardia 10 (1), 1-16.

Wood, G.A.R. and Lass, R.A. (2001) Cacao, 4th edn. Oxford: Blackwell Science Ltd. 620 pp.

Zeilinger, S. and Omann, M. (2007). *Trichoderma* Biocontrol: Signal Transduction Pathways Involved in Host Sensing and Mycoparasitism. Gene Regulation and Systems Biology (1), 227-234.

Health and Safety Considerations   **2**

Abstract

Standard health and safety protocols are an important requirement in *Trichoderma* production and nursery screening activities, both in the laboratory, nursery and the field. The standards may vary depending on the requirements needed. Identification and elimination of hazards and risks in some activities, followed by developing specific safety procedures, and responding immediately to workplace accidents, and injuries, are important features in establishing an effective occupational health and safety programme. Protocols of health and safety issues relating to *Trichoderma* production, nursery screening activities, *Trichoderma* storage and field activities for oil palm commodities are explained below.

2.1   Health and Safety in the Laboratory

Producing both *Trichoderma* and *Ganoderma* inoculum consist of many activities and procedures. Those activities consist of isolating and culturing fungal isolates, preparing the inoculum, propagating the inoculums and field application. Therefore, good laboratory, nursery and field practices are recommended. Labour safety has also become the important consideration.

- Clean and wash hands before entering the laboratory. This will give protection for the worker and samples from outside contamination.
- Laboratory coats and shoes must be worn when entering the laboratory and removed when leaving the laboratory. This will give protection for the workers and the samples from outside contamination and also people outside from laboratory contamination.
- Face masks are required when working with ethanol and contaminated samples. This is good for the operators and for reducing sample contamination.

- Gloves are required when working with hazardous chemicals, hot, heavy and hard materials.
- Wear appropriate clothing, skin exposure should be minimized. Field clothing and Wellington boots should be removed before entering the laboratory.
- Training in inoculating and flaming techniques.
- Training in operating an autoclave, oven sterilizer, and other equipment used in the lab.
- Be aware of emergency procedures: fire-fighting, emergency exits, emergency phone numbers, and the location of fire extinguishers, emergency shower, eye-wash and first aid/first aiders.
- Be aware of hazards relating to chemicals used in the laboratory and their Material Safety Data Sheet (MSDS information is available on the web), which provide information on health and safety, first aid, fire, explosion risks, disposal, how to clean up spillage, handling and storage.
- Be aware of standard operating procedures (SOPs) that have been developed for your laboratory, or which should be developed, such as for waste disposal, e.g. plastic bags.
- Check as soon as possible for any broken laboratory equipment which may be dangerous for the worker.

2.2   Health and Safety in the Nursery

Activities in the nursery are simpler than in the field. Treatment for the experiment is also done when the trial is started, therefore daily activities are only watering and cleaning the area, then the weekly or monthly activity is recording. However, attention should be paid to the health and safety of the workers as well as to the environment.

Equipment required in the *Trichoderma* and *Ganoderma* seedling screening nursery are:

- Sieve used to sift soil
- Hoe used to hoe the soil to be sifted
- Sharp knife used to open rubber wood block (RWB) and *Trichoderma* package
- Wheelbarrow used to transport soil and polybags inside nursery

Safety equipment which needs to be worn in the nursery:

- Rubber boots
- Gloves for handling sharp or hard materials
- Masks if working with *Trichoderma* powder or chemicals such as pesticide
- Raincoat during the rainy season
- Hats

Training and refresh training are essential for worker safety. Other considerations are:

- Chemical use (especially pesticides)
- Standard Operating Procedures (SOPs)
- Working alone
- Emergency procedures, first aid box
- Be aware of nuisance insects and other animals (e.g. monkey and snakes)

2.3   Health and Safety in the Field

Field activities are generally done on a large scale. Trialling is usually conducted with at least one block of 10-20 ha areas, and *Trichoderma* application in replanting programmes is normally done for a minimum of one block of 20 hectares. Attention should be paid to the environment and the health and safety of the workers.

Equipment required in the *Trichoderma* and *Ganoderma* seedling screening and commercial application during replanting in the field are:

- Spade used to dig the soil
- Sharp knife used to untie the seedlings after planting in the planting hole
- Bucket used to transport biopesticide products (if any)

Safety equipment which needs to be worn in the field:

- Rubber boots
- Gloves for handling sharp or hard materials
- Masks if working with *Trichoderma* powder or chemicals such as pesticide
- Raincoat during the rainy season
- Hats

Training and refresher training are essential for worker safety. Other considerations are:

- Chemical use (especially pesticides)
- Standard Operating Procedures (SOPs)
- Working alone
- Emergency procedures, first aid box
- Be aware of nuisance insects and other animals (e.g. monkey and snakes)

2.4   On the Safety of *Trichoderma* spp. in Biocontrol Applications

Most filamentous fungi, including *Trichoderma* spp. are capable of producing secondary metabolites (Mukherjee *et al.*, 2012). Hence when working with *Trichoderma*, general caution and careful consideration of clean work and appropriate training is advised. Moreover, strains of the genus *Trichoderma* were also found in clinical samples, particularly in immunocompromised patients (Kredics *et al.*, 2003). *Trichoderma* spp. have also been mentioned as emerging and uncommon human pathogens (Walsh *et al.*, 2004). Although there are many beneficial properties and application of various *Trichoderma* species (Mukherjee *et al.*, 2012; 2013) there are a growing number of disease incidences caused by various *Trichoderma* species in human (Hatvani *et al.*, 2013). Most cases are associated with compromised immune systems, but other people at risk include dialysis patients, and people with leukaemia, blood disorders and HIV. The species involved include: *T. viride*, *T. longibrachiatum*, *T. citrinoviride*, *T. koningii*, *T. pseudokoningii*, *T. harzianum*, *T. reesei* and *T. altoviride*. Most belong to the Longibrachiatum section of the *Trichoderma* genus. Infections are thought to be primarily air borne and early symptoms are often respiratory problems. As a human pathogen these species must be able to grow at 37°C. Treatments usually involve the use of anti-fungal drugs. Generally, *Trichoderma* spp. are considered opportunistic pathogens which usually grow in soil and on decaying plant material. Some of the species listed above, *T. viride*, *T. koningii* and *T. harzianum* have been investigated in the control of Ganoderma. *T. virens* is the subject of this manual, it does not belong to the Longibrachiatum section and is unable to grow at 37°C. Note, *Trichoderma* spp. are ubiquitous in our environment and many of the species mentioned above can be found in the environment, from tropical to temperate regions. Additionally, *Trichoderma* spp. have a long history of safe use in biocontrol, also in developing countries and for diverse applications. Therefore, the overall health risk can be considered acceptable. Nevertheless, it is strongly advised that people working with *Trichoderma* do not belong to any risk group and workers should seek immediate medical attention if they become sick or show symptoms related to *Trichoderma* infection.

Bibliography

Anonym (2012) Health and Safety for Greenhouses and Nurseries. Workers' Compensation Board of British Columbia, Canada. 120p.

Barker, K. (2005) At the bench: a laboratory navigator. Cold Spring Harbor Press, New York, USA.

Hatvani, L., Manczinger, L., Vagvolgyi, C. and Kredits, L. (2013) *Trichoderma* as human pathogen. In Mukherjee PK, Horwitz BA, Singh US, Mukherjee M, Schmoll M (eds) *Trichoderma* Biology and Applications. CAB International Wallingford pp 292-313.

Kredics, L., Antal, Z., Manczinger, L., Kevei, F. and Nagy, E. (2003) Clinical Importance of The Genus *Trichoderma*. A review. Acta Microbiology and Immunology, Hungary. 50, 105-117.

Mukherjee, P. K., Horwitz, B. A. and Kenerley, C. M. (2012) Secondary metabolism in Trichoderma – a genomic perspective. Microbiology 158, 35–45.

Mukherjee, P.K., Horwitz, B.A., Singh, U.S., Mukherjee, M. and Schmoll, M. (2013) *Trichoderma* Biology and Applications. CAB International Wallingford pp 292-313.

Walsh, T.J., Groll. A., Hiemenz, J., Fleming. R., Roilides, E. and Anaissie, E. (2004) Infections due to emerging and uncommon medically important fungal pathogens. Clinical Microbiology and Infection 10, Supplement 1, 48-66.

Culturing *Trichoderma*

3

Abstract

The first step in propagating *Trichoderma* in culture (*in vitro*) is the preparation of growth media. This chapter explains how to prepare the different media used in culturing *Trichoderma* isolates used by Verdant. To propagate *Trichoderma* isolates for use in integrated pest management of *Ganoderma* diseases of oil palm the usual medium used is Potato Sucrose Agar (PSA), followed by commercial propagation in a maize medium, and further propagation, in rice husk medium. This is then branded and sold for soil application to control *Ganoderma* in plantations.

3.1 Potato Sucrose Agar (PSA) Medium Preparation

Potato Sucrose agar (PSA) is one of the most commonly used media for the isolation and cultivation of fungi. The methods provided here for PSA preparation are for one litre of medium (Table 3.1). This may be scaled up or down depending on capacity and volumes required.

Table 3.1. Materials needed to prepare 1 litre of PSA medium

Materials	Quantity
Potato	200 g
Agar-agar powder	15-25 g (according to the brand)
Sucrose	2 g
Aquadest (purified water)	1 litre
Ethanol 70%	As necessary (used for surface sterilization)

Equipment and tools needed for media preparation

a. Gloves, workers should wear gloves to prevent contamination of their skin and materials

b. Knife, used to slice potatoes

c. Graduated jug to dissolve sucrose, agar and other components before making up to 1 litre

d. Analytical balance, used for weighing components of the medium

e. Boiling pan, used to boil the potato extract

f. Stove/Hot plate, used to boil potato pieces and make PSA extract

g. Sieve, used to separate potato pulp from the extract

h. Measuring cylinder (1 litre and 40 ml), used to place ingredients such as agar-agar powder and sucrose (1 litre) and make up required volume; potato extract is poured into 40 ml bottles

i. Stirring hot plate, used to stir and boil the PSA extract

j. Clean glass bottles (40 ml), used for culturing

k. Funnel, used to pour PSA extract from measuring glass bottles

l. Flat-sided glass bottles, used to store the PSA extract

m. Stoppers (cotton wool and paper/newspaper), used to close the glass bottles containing PSA extract

n. Thread, or rubber bands, used to tie paper to glass bottles

o. Autoclave, used for media and equipment sterilization

p. Laminar flow cabinet, used for sterile operations, e.g. placement of PSA slope media

PSA medium is commonly used to propagate or culture *Trichoderma*. The growth of the fungus is generally more rapid and denser on PSA than other media. This is because PSA contains abundant carbohydrates (from potato and sucrose). In general, *Trichoderma* is rarely contaminated by other microorganisms, therefore PSA requires less antibiotics and this also allows the fungus to grow rapidly and densely.

Steps in preparing one litre of PSA medium

Step 1
Before starting, all equipment and work places must be checked. Prepare all materials needed based on requirements. Potatoes should be peeled before they are weighed.

Fig. 3.1. Potatoes are peeled before being weighed

Step 2
Cut the potatoes into cubes of about 2-3 cm^3

Fig. 3.2. Peeling and cutting potatoes

Step 3
To prepare potato extract, boil 200 g cubed potato in 1 litre distilled water for about 20 minutes (± 5 minutes). The boiled potato pieces are cooled and sieved before being mixed with the other ingredients.

Fig. 3.3. Boiling potatoes

Step 4
Other materials (agar powder and sucrose) are mixed in a graduated jug and aquadest (sterile water) added to dissolve materials. Then add the cooled potato extract and stir manually. Boil again on a hot plate for 10 minutes (± 5 minutes).

Fig. 3.4. Agar powder and sucrose are mixed in a graduated jug before cooled PSA extract is added and the volume made up to 1 litre

Fig. 3.5. Potato extract is poured through a sieve and the volume measured in a graduated jug

Fig. 3.6. Potato extract is poured into the pan, stirred manually and boiled

Step 5
The medium is then poured into bottles (40 ml per bottle). Glass bottles are prepared together with their stoppers (consisting of non-absorbent cotton wool bungs) and paper is used to cover the stopper and tied down with thread or rubber band.

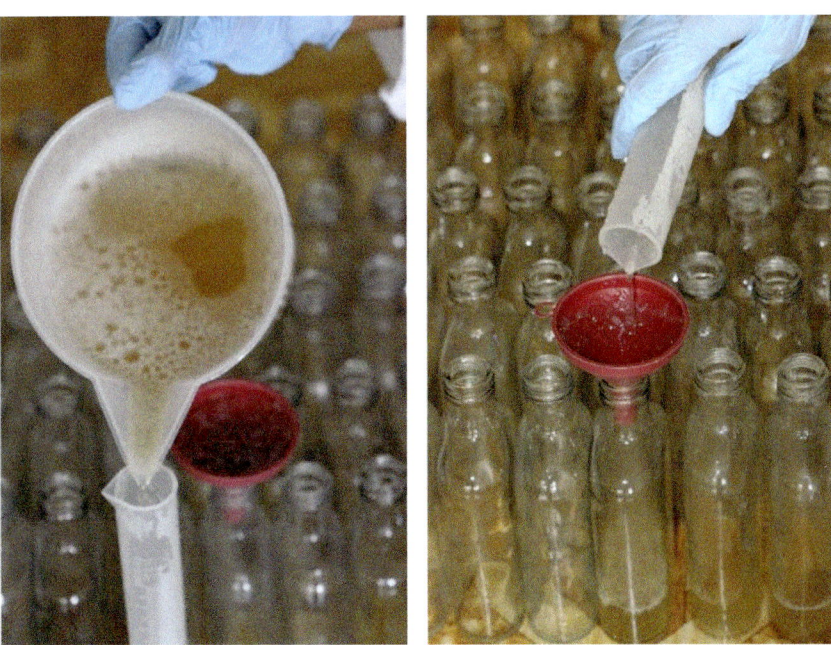

Fig. 3.7. The medium is poured into clean glass bottles

Step 6
Sterilization of the potato medium is conducted by autoclaving at 0.11 MPa pressure and at 120°C for 1 hour. Both horizontal and vertical autoclaves

may be used to sterilize the medium. Since the autoclave uses high pressure and high temperature, operating instructions for operating specific autoclaves must be strictly adhered to and training should be given to the workers for the specific models used.

Fig. 3.8. Placement of bottles (containing PSA media and sealed with cotton wool bungs and newspaper) inside an autoclave

Step 7
Clean all work surfaces of laminar cabinet with a disinfectant solution before PSA media bottles are placed inside. The PSA media are then sloped by placing the media bottles inside the laminar air flow cabinet at an angle of 30 degrees, where they are cooled and allowed to set, to be used the next day.

Fig. 3.9. Placement (at angle of 30 degrees) of bottles containing PSA media in laminar air flow cabinet

3.2 Maize Medium Preparation

Maize (also called corn) is a good medium to grow *Trichoderma* because it contains carbohydrates to support fungal growth. The advantages of maize over rice husk (see Section 3.3) is it has a larger surface area which allows the fungus to grow and produce more spores. The medium is made up simply of sterilized maize.

Equipment and tools needed for media preparation

a. Plastic basket, used to wash milled maize
b. Balance to weigh out 500 g of milled maize
c. Autoclavable plastic bags, used for packaging the maize
d. Non-absorbent cotton wool and paper/newspaper, used to make stoppers
e. PVC pipe with diameter ¾ inch, used inside plastic bags with milled maize to make bag closure easy
f. Yarn, used to bind the PVC pipe in the plastic bag
g. Rubber bands, used to tie up the paper to close the stopper
h. Autoclave, used to sterilize the milled maize in the plastic bags
i. Plastic basket, used to place the milled maize after sterilization

Steps in making maize media

Step 1
Wash milled maize 3 to 4 times with water until the water runs clear. Then place about 500 g of the milled maize into an autoclavable plastic bag. The plastic bag is prepared together with its stopper (consisting of non-absorbent cotton wool and paper/newspaper). Place the PVC pipe (with a diameter ¾ inch) in the mouth of the plastic bag and bind it by using yarn, then close it with a stopper.

Fig. 3.10. Milled maize packaged in autoclavable plastic bags

Step 2
Sterilize the maize medium using an autoclave at (120–121°C) for 1 hour. Both horizontal and vertical autoclaves may be used.

Fig. 3.11. Milled maize medium sterilization by using vertical autoclave

Step 3

After sterilization, take the maize media bags from the autoclave and place them into a plastic basket and allow to cool.

Fig. 3.12. Sterilized maize in cooling basket

Step 4

Fungal inoculation is normally done the next day. Details of the inoculation are given in Chapter 4.2. *Trichoderma* multiplication on maize medium.

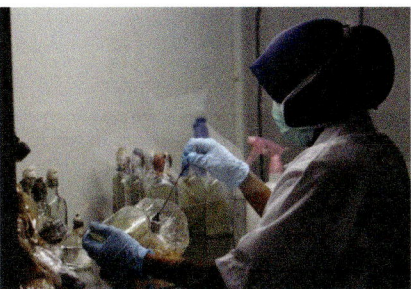 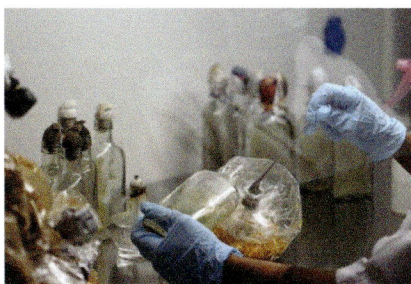

Fig. 3.13. Fungal inoculation on maize media

3.3 Rice Husk Medium Preparation

Rice husk medium is used for field application of *Trichoderma*. *Trichoderma* is propagated on maize corn media and then this is used to inoculate rice husk medium. Rice husk medium is suitable to grow *Trichoderma* because it contains carbohydrates suitable for the fungus to grow. Therefore, *Trichoderma* grows quickly on rice husk medium. Also, *Trichoderma* survives for a relatively long time (up to 8 months) on rice husk medium. For large scale production, rice husk medium can be sterilized using sun drying or a large, modified oven.

Equipment and tools required

a. Permanent area (6 m x 6 m is normally used for 14 tons), cement floor or tarpaulins, used to dry the rice husk using sun drying
b. Shovel, to turn rice husk manually
c. Mask and rubber gloves are used by workers
d. Gunny bags are used to store the rice husk before and after sun drying

Steps in making culture medium

Step 1

Prepare a large (6 m x 6 m) permanent area to be used to dry the rice husk under the sun.

Fig. 3.14. Permanent area for rice husk sun drying

Step 2
Place the husk on the cement floor or tarpaulins (when grassed area is used) to dry.

Fig. 3.15. Rice husk placement in sun drying area

Step 3
The rice husk is manually turned every hour using a shovel to speed up drying.

Step 4
On a fairly sunny day (in tropical regions), 7 hours of sun drying is usually optimal, typical hours are from 7 a.m. to 2 p.m.

To dry about 400 kg per day, two labourers are required. Face masks and rubber gloves should be used by the labourers.

The drying also serves to partially sterilize the rice husk, which is sufficient for small scale production.

Step 5
After the rice husk is dry, it is packed into gunny bags.

Fig. 3.16. Packing of rice husk to gunny bag

Bibliography

Gill, S.V., Pastor, S. and March, G.J. (2009) Quantitative isolation of biocontrol agents *Trichoderma* spp., *Gliocladium* spp. and actinomycetes from soil with culture media. *Microbiological Research* 164, 196-205.

Hauser, J.T. (2006) *Techniques for Studying Bacteria and Fungi*. Carolina Biological Supply Company, Burlington, USA, 31pp.

Himedia (2016) *Potato Dextrose Agar*. Technical Data. Himedia Laboratories, Mumbai, India.

Nevalainen, H., Te'o, V.S.J. and Kautto, L. (2014) Methods for isolation and cultivation of filamentous fungi. In: Clifton, N.J. (Eds) *Methods in Molecular Biology*. Pubmed, pp. 1-16.

Trichoderma Multiplication

4

Abstract

The mass production of *Trichoderma* is described. The fungus is propagated on potato sucrose agar and maize media. This chapter explains how to propagate *Trichoderma* on these media.

4.1   *Trichoderma* Multiplication on PSA

PSA is the preferred media for mass production of *Trichoderma* cultures. *Trichoderma* grows fast on this media and produces more biomass per time than other media e.g. TSM (*Trichoderma* selective media). In addition, the cost for large scale production is relatively less than other media.

Equipment and tools needed for *Trichoderma* multiplication

a. Glass bottles or Petri dishes, used to culture pure *Trichoderma* isolates

b. Innoculation loop, used to transfer pure culture pieces (full of spore in the loop) of *Trichoderma* spore to PSA medium

c. A spirit lamp or an electric hot glass bead sterilizer, used to sterilize transfer tools

d. 70% ethanol, used for surface disinfection of working areas and tools

e. Stoppers (cotton wool bungs or gauze and newspaper), used to close the glass bottle

f. Thread or rubber bands, used to tie paper over bungs

g. Tape, used to seal culture containers

h. Cell tape or parafilm, used to seal Petri dishes

i. Laminar air flow cabinet used to maintain sterile conditions for operations

j. Marker pen, used for labelling

Steps in *Trichoderma* multiplication on PSA

Step 1
Prepare the PSA medium in glass bottles or Petri dishes and inoculate with a pure *Trichoderma* isolate. Bottles are normally recommended because they provide a larger surface area for the fungus to grow. Before starting to culture the selected isolate, all the surfaces of the equipment and tools must be cleaned with a disinfectant solution or 70% ethanol. Transfer needles should be flame sterilized using a Bunsen burner, spirit lamp or an electric hot glass bead sterilizer. Inoculation is performed inside a laminar flow cabinet.

Fig. 4.1. Equipment used for *Trichoderma* multiplication: from left to right: *Trichoderma* pure isolate, PSA medium in glass bottle, spirit lamp, transfer needle, 70% ethanol

Step 2
Remove a portion of the *Trichoderma* culture that is to be reproduced from the mature culture in a glass bottle, using an inoculating loop. and inoculate onto the fresh PSA medium.

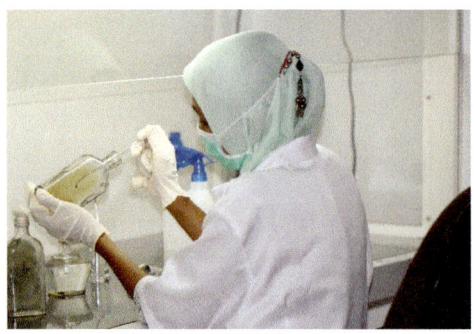

Fig. 4.2. Transfer of *Trichoderma* multiplication from pure isolate to PSA medium

Step 3
Close the glass bottle using a stopper (consisting of non-absorbent cotton bung or absorbent gauze and paper/newspaper) and strengthen the Petri dish by sealing with tape or Parafilm.

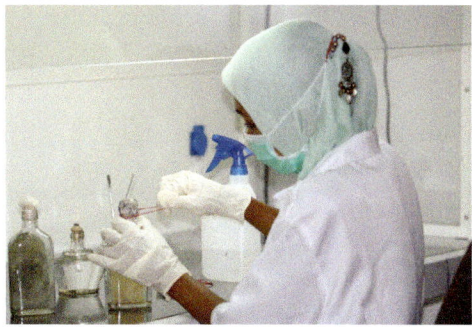

Fig. 4.3. *Trichoderma* on PSA medium in glass bottle

Step 4
Label: write the name of the isolate and sub-culturing date on the culture vessel. The medium can be stored at 25–30°C. Wait for 7–10 days until mycelium covers the surface, it is then ready to be used to inoculate maize medium in bags.

Fig. 4.4. Incubation of *Trichoderma* in PSA medium

4.2   *Trichoderma* Multiplication on Maize Medium

One day after the sterilization process, maize medium can be used to propagate *Trichoderma* isolate from sloped culture bottles.

Materials, equipment and tools

1. Sporulating *Trichoderma* culture from PSA in glass bottles
2. Spatula spoon, used to transfer *Trichoderma* from glass bottles to maize medium
3. 70% ethanol or other disinfectant, to sterilize work surfaces and tools
4. A spirit lamp or hot glass beads, to sterilize tools
5. Laminar air flow cabinet is for all sterile work

6. Autoclavable plastic bags, used to package the maize medium
7. Stoppers (cotton wool and newspaper), used to seal the glass media bottles
8. Marker pen, used for labelling

Steps in *Trichoderma* multiplication on maize medium

Step 1
Prepare equipment and tools. Before starting to culture the isolate, all the work surfaces, equipment and tools must be cleaned with a disinfectant solution or 70% ethanol. The spatula spoon is flame sterilized using a spirit lamp or hot glass beads. Inoculation is done inside the laminar flow cabinet.

Step 2
Cut the pure isolate in bottled media into small pieces (5x5 mm) by using a spatula spoon, take some slices of the isolate, and place onto the maize medium. One bottle of the *Trichoderma* isolate can be used for three packs of maize medium. The isolate should be placed evenly over the maize medium to promote faster *Trichoderma* growth.

Fig. 4.5. Process of *Trichoderma* multiplication from pure isolate to corn

Step 3
Close and strengthen the plastic bag containing media using a stopper (consisting of absorbent cotton wool and paper/newspaper).

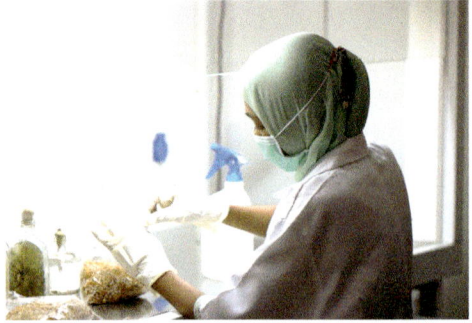

Fig. 4.6. Preparation of media bags

Step 4

Label: write the name of the isolate and sub-culturing date. The medium can be stored at 25–30°C. Wait for 17 to 20 days until mycelium covers the surface and is ready to be used.

Fig. 4.7. Incubation of the labelled cultures

Bibliography

Kumar, S., Thakur, M. and Rani, A. (2014) *Trichoderma*: Mass production, formulation, quality control, delivery and its scope in commercialization in India for the management of plant diseases. *Academic Journals* 9(53), 3838-3852.

Rahman, M.A., Begum, M.F. and Alam, M.F. (2009) Screening of *Trichoderma* isolates as a biological control agent against *Ceratocystis paradoxa* causing pineapple disease of sugarcane. *The Korean Society of Mycology* 37(4), 277-285.

Ganoderma Pathogenicity Testing **5**

Abstract

Trichoderma isolates may be screened for their effectiveness in antagonizing *Ganoderma* in both nursery and field tests. For this, *Ganoderma* as a source of disease (inoculum) is required. A recommended first step is to collect *Ganoderma* basidiocarps from *Ganoderma* infected palms and to then culture these on an appropriate medium. After that, the cultures can be maintained and used as a source of inoculum for nursery seedling screening using *Ganoderma* inoculated rubber wood blocks (RWBs) as a disease source. This chapter describes how to collect and culture *Ganoderma*, and how to produce *Ganoderma* infected rubber wood blocks to be used as an infection source in nursery screening.

5.1   Collection of *Ganoderma* Basidiocarps

Basidiocarps may be isolated from any part of a *Ganoderma* diseased oil palm in the field. Most *Ganoderma* infected palms have basidiocarps on the stem, which are easy to collect using a knife. However, for Upper Stem Rot (USR) basidiocarps are positioned higher and a chainsaw may be required to fell the upper parts of the palm to collect the basidiocarps. The use of young, actively growing basidiocarps is preferred as they have more actively growing mycelium than older ones. Also, the old basidiocarps have very thin mycelia which are difficult to isolate. The difference between young and old basidiocarps can be seen by their colour, old basidiocarps are usually darker in colour than the young ones (Fig. 5.1). Moreover, the upper surface of old basidiocarps is harder and more difficult to work with.

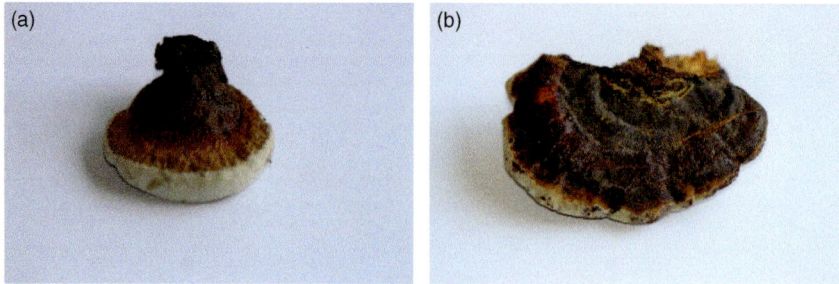

Fig. 5.1. a) Young, actively growing basidiocarp and b) old basidiocarp

Table 5.1. Equipment and tools needed for basidiocarp collection

Materials
Paper bag for samples
Ganoderma basidiocarp
Gloves
Marker
Knife
Chainsaw

Steps in basidiocarp collection

Step 1

Before starting, the preparation of all equipment and work places must be checked for operational use. Select a *Ganoderma* diseased palm to be sampled. Prepare all equipment needed based on requirements (Table 5.1), then go to the field, and make sure it's appropriate.

Fig. 5.2. Example of a *Ganoderma* diseased palm

Step 2

Ganoderma basidiocarps are harvested from infected palms (BSR or USR palms). BSR basidiocarps are located on the palm trunk base and can be harvested using a knife. Basidiocarps of USR are located in the middle or top of the palm and harvested using a chain saw to fell the palm and then basidiocarps can be removed using a knife. The chain saw should only be used by a trained worker with appropriate safety precautions taken (see Chapter 2).

Fig. 5.3. Young, actively growing *Ganoderma* basidiocarps on a diseased palm

Step 4
Place the detached basidiocarp sample inside the paper bag, one sample per bag.

Step 5
Take notes: record date, estate name, field/block number, row and palm number, disease (*Ganoderma*) score of the sampled palm, collector's name.

Fig. 5.4. A collection bag for basidiocarp sample

Samples taken from the field are usually cultured to the medium directly in the same day. However, if the sample is too much, then the rest of the sample with the collection bag is kept in the refrigerator. The basidiocarp is usually kept for up to four days.

5.2   *Ganoderma* Culture Preparation

Ganoderma cultures may be obtained from the basidiocarps (a basidiocarp is the fruiting body of basidiomycete fungi) taken from infected palm tissues (trunk or roots) or from basidiospores of *Ganoderma* collected from infected palms in the field. However, the isolation method using basidiocarps is much easier as collecting spores is difficult. Therefore, it is the only method that will be explained in this chapter. An aseptic laboratory environment is needed for successful isolation and culture of *Ganoderma*.

Water Agar (WA) or *Ganoderma* Selective Medium (GSM) may be used as the first culture medium, later modified Potato Sucrose Agar (PSA), plus antibiotics, is used for rapid multiplication of *Ganoderma* cultures. The selection of samples especially from the basidiocarp can affect the success of *Ganoderma* isolation so, a vigorous, actively growing basidiocarp is optimal as the initial sample.

Table 5.2. Materials needed for media preparation (PSA plus antibiotics)

Materials	Quantity (g/ml)
Potato	200 g
Ganoderma basidiocarp	
Agar-agar bars	14 g
Sucrose	2 g
Chloramphenicol	0.1 g
Streptomicyn	0.3 g
Aquadest	1 litre
Ethanol 70%	As necessary

Table 5.3. Check list of equipment and tools needed for media preparation (PSA with antibiotics)

Equipment and tools	PSA plus antibiotics
Autoclave	√
Analytical balance	√
Measuring glass	√
Erlenmeyer flask 2 L	√
Boiling pan	√
Sieve	√
Stirring hot plate	√
Stove/hot plate	√
Aluminium foil	√
A spirit lamp	√
Laminar flow cabinet	√
Glass Petri dish	√

Steps in preparing one litre of PSA plus antibiotics medium

Step 1
Before starting the preparation, all equipment and work places must be checked for operational use. Prepare all materials needed based on requirements. Potatoes should be peeled before being weighed.

Step 2
To prepare potato extract, boil 200 g sliced potato in 1 litre distilled water for about 20 minutes (± 5 minutes). The boiled potato pieces are cooled for about 10 minutes and sieved before being mixed with the other ingredients (Table 5.2).

Step 3

Other materials (agar powder, sucrose, streptomycin sulphate, and chlor-
amphenicol) are first mixed in an Erlenmeyer flask, the potato extract is
poured into the Erlenmeyer flask (which is covered with aluminium foil),
then stirred using a stirring hot plate (Table 5.3) for about 10 minutes.

Fig. 5.5. Mixed potato extract
and other materials in an
Erlenmeyer flask

Step 4

Sterilization of the medium is conducted by autoclaving at 0.11 MPa pressure (120–121°C) for 1 hour. Vertical autoclaves can be used to sterilize the
medium. Since autoclaves use high pressure and high temperatures, work instructions for operating autoclaves must be strictly adhered to and training
must be given to the workers prior to any use. Note: different autoclaves
require different procedures.

Step 5

After sterilization, cool the medium on a stirrer hot plate for one and a half
hours before it is poured into Petri dishes.

Step 6

Clean all work surfaces of the laminar flow cabinet with a disinfectant solution before the PSA is placed inside. Pour the PSA plus antibiotics into Petri
dishes (at a temperature of about 35°C) inside a laminar air flow cabinet.
The poured plates are then allowed to cool and stored in the laminar air
flow to be used next day.

Fig. 5.6. Pouring PSA plus antibiotics medium into Petri dishes inside a laminar air flow cabinet

Next day: pieces of the sampled basidiocarps can be plated onto the PSA plus antibiotic medium. The materials for producing cultures from the basidiocarps are given below:

Table 5.4. Materials needed for initiation of *Ganoderma* cultures (from basidiocarp samples) onto PSA medium plus antibiotics.

Materials	Quantity
Ganoderma basidiocarp	As necessary
PSA with antibiotics	As necessary
Knife	1 pc
Ethanol 70%	As necessary
Cutting board	1 pc
Glass beaker	2 pcs
A spirit lamp	1 pcs
Cell tape or Parafilm	As necessary
Forceps	1 pc
Inoculation blade	1 pc
Laminar flow cabinet	1 unit
Sterile Petri dishes	As necessary
Marker pen	1 unit

Steps in *Ganoderma* culture from basidiocarp to PSA medium

Step 1
In the laboratory, samples taken from the basidiocarps collected in the field are cut into small pieces (1 cm x 1 cm x 1 cm). The basidiocarp should be cut transversely and samples of the tissue removed from inside the soft fleshy part of the structure (Fig. 5.7). The pieces of tissue are surface sterilized by washing the flesh pieces with water and then soaking in 70% ethanol for approximately 15–20 seconds.

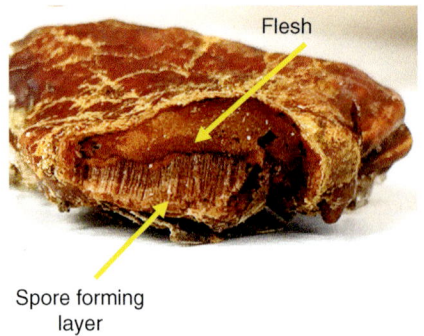

Fig. 5.7. *Ganoderma* basidiocarp consists of flesh and spore forming layer

Flesh

Spore forming layer

Step 2
These larger pieces are then cut again into smaller pieces (5x5x5 mm), and each piece placed onto solid PSA plus antibiotics (4-5 pieces per plate). The work is done inside a laminar air flow cabinet.

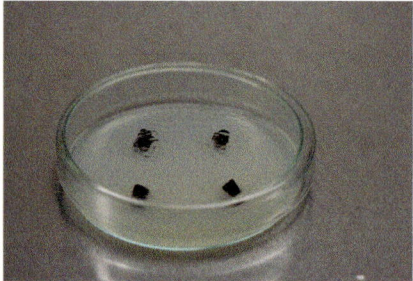

Fig. 5.8. Small pieces of basidiocarp on PSA plus antibiotics medium

Step 3
Use cell tape or Parafilm to seal the edge of the Petri dish to avoid contamination.

Step 4
Label: write the isolate's name, isolation date, sample code and medium's name on the surface of the plate.

Step 5
Observe samples daily until mycelium is seen growing around the plated samples. The length of time to see mycelial growth is variable and depends on the sample. Samples from older basidiocarps will take longer to produce mycelium than the younger ones. Normally, this process takes 2–4 weeks from the initial culture of young, actively growing basidiocarp samples. Using a sterile blade, take a piece of the culture from the edge of the culture. Select the

most actively growing areas of mycelial growth and that which is furthest away from the original sample as this will lessen chances of contamination. Place the inoculum onto fresh PSA plus antibiotic medium. This process can be repeated until cultures are obtained that are free from contamination. All the fungi are incubated under room temperature (27–30°C) in the dark. For long-term storage or conservation, pieces of culture (solid medium containing mycelium) are placed in a flask with sterile pure water and stored in an incubator at 20°C. Cultures may be kept for more than one year.

5.3   *Ganoderma* Multiplication in Culture

The use of PSA plus antibiotics is preferred to propagate *Ganoderma* isolates, because the fungus produces more biomass on this medium and the antibiotics reduce bacterial contamination.

Table 5.5. Materials, equipment and tools needed for *Ganoderma* propagation in culture

Materials	Equipment and Tools
Ganoderma pure culture PSA media	A methylated spirit lamp Cell tape or Parafilm Inoculation blade Laminar flow cabinet Marker pen Sterile Petri dishes

Steps in multiplying *Ganoderma* in culture

Step 1
Prepare both the PSA plus antibiotic medium and obtain pure *Ganoderma* isolate (see Table 5.5 above). Before proceeding, all the surfaces of the equipment and tools must be cleaned with a disinfectant solution or 70% ethanol. The inoculation blade is flame-sterilized using a spirit lamp.

Step 2
Cut the pure culture into small pieces (5x5 mm), take one slice of each culture, and place (mycelium facing down) onto fresh PSA medium.

Step 3
Close and seal the Petri dish with tape or Parafilm.

Step 4
Label: write the name of the isolate, sub-culturing date and name of the medium on the surface of the Petri dishes. The medium can be stored at 27–30°C in the dark.

Wait for 2 weeks to 1 month until the mycelia cover the surface of the plate and it is ready to be used.

5.4   Preparation of Rubber Wood Blocks

Rubber wood blocks (RWBs) are used as a *Ganoderma* source for seedling screening (Rahmaningsih *et al.*, 2018, this series). The use of RWBs has been observed to be the most effective substrate for *Ganoderma*: studies comparing RWBs with other substrates such as liquid medium, sawdust-based and oil palm wood blocks, showed RWBs to be the most effective (Breton *et al.*, 2006). Moreover, the availability of rubber wood is consistent since there are often many rubber plantations in areas where oil palm is grown. The unproductive rubber wood from the field is usually sent to small local saw mills to produce wood blocks. The size of the RWB can vary, here we use 6x6x6 cm and 6x6x3 cm blocks. However, there is a problem with RWB waste after use. Since zero burning systems are applied in all oil palm plantations, other methods of destroying the *Ganoderma* inoculum waste should be considered e.g. incineration.

Potato agar (PA) medium is used as a starter for *Ganoderma* to grow prior to growth on the RWBs for inoculation. This medium is applied to the RWBs. This agar medium differs from PSA plus antibiotics as it does not contain antibiotics (pure cultures are being used and RWBs are less likely to become contaminated) and PA is cheaper for mass multiplication of *Ganoderma*. The materials and equipment used to make PA are noted below.

Table 5.6. Materials, equipment and tools needed for PA preparation

Materials (for one litre of PA media)	Equipment and Tools
Agar-agar powder 15 g	Analytical balance
Sucrose 2 g	Autoclave 75X/big pan (size 48 cm x 39 cm)
Ethanol 70% as necessary	Boiling pan
Distilled water 1.5 L	Knife
Potatoes 200 g	Measuring glass cylinder
	Sieve
	Stove
	Pail
	Stirrer hot Plate

Note: PA medium is made a day after RWB preparation. PA is the same as PSA but without antibiotics.

Steps in the preparation of rubber wood blocks

Step 1
Prepare all materials needed based on requirements. Potatoes should be peeled before being weighed. Cut the potatoes into small pieces (1 cm³) and boil them for 20 minutes (± 5 minutes). Time is counted when the water begins to boil.

Step 2
Sieve the potato extract and measure the extract amount based on the requirements. Boil the extract again with agar-agar powder and sugar.

Step 3
The medium should be poured directly onto the RWBs whilst hot (±75°C).

Table 5.7. Materials, equipment and tools needed for the preparation of RWBs

Materials	Equipment and Tools
PA medium (50 cc for each bag of RWBs) RWB size 6 x 6 x 6 cm or 6 x 6 x 3 cm	Aluminium foil/newspaper Autoclave Boiling pan (size 48 cm x 39 cm) Marker pen Measuring glass cylinder Polypropylene plastic bag PVC pipe 1½ inch Stopper from hydrophilic cotton and gauze Stove Wool thread

Step 4
Prepare RWBs based on the requirements needed. Soak rubber wood in water inside a large (25 litre) pail (70 cm diameter) overnight to maximize imbibition.

Fig. 5.9. Rubber wood soaked in water inside a big pail

Step 5
Boil the RWBs for 5–7 hours. Time is counted when the water begins to boil.
Top up water to ensure the RWBs are covered by boiling water.

Fig. 5.10. Rubber wood blocks
being boiled

Step 6
Next working day: pack the RWBs in plastic bags (two RWBs per polybag).

Step 7
Put the mouth of the plastic bag into the PVC pipe and bind it with thread.
Each bag is filled with 2 RWBs.

Fig. 5.11. Rubber wood blocks
placed in plastic bag

Step 8
Fill each bag with 50 ml (±2 ml) of PA medium (Step 3). Measure the medium
with a plastic cylinder.

Step 9
Plug the mouth of the plastic bag with a stopper and then cover it with aluminium foil or newspaper.

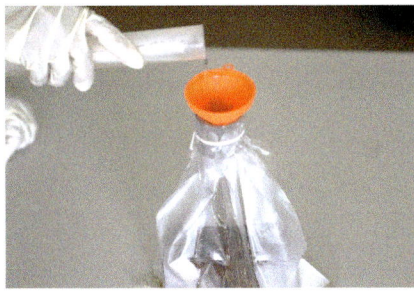 

Fig. 5.12. Rubber wood and PA medium placed in plastic bag and closed with stopper

Step 10
Sterilize the RWBs (plus PA) by autoclaving at 0.11 MPa pressure (120–121°C) for one hour. Manual boiling is an alternative to autoclaving: manual boiler (big pan, volume 200 litre) can be used with 5 hours of boiling.

Fig. 5.13. Sterilize RWBs and PA medium using an autoclave

Step 11
After the sterilization (boiling) process, allow the RWBs to cool to 27°C.

Step 12
Leave the RWBs for one night and inoculate them the next day.

5.5   Inoculation of Rubber Wood Blocks

Ganoderma inoculation of RWBs can be done one day after RWB sterilization. This process is similar to other culture preparations. The only difference is the medium used. Culture preparation uses PSA plus antibiotics as the medium for *Ganoderma* to grow while this process uses RWBs as a substrate and PA medium as a starter. Once the *Ganoderma* isolates have grown well on the RWB media, the pathogenicity of the isolates can be tested on seedlings in the nursery (Rahmaningsih *et al.*, 2018, this series). The most aggressive isolates may be genotyped by DNA markers and used as a standard source inoculum in the nursery. This can allow comparisons between trials. For example, an investigation by Rakib *et al.* (2015) found that an isolate of *G. zonatum* of USR was the most aggressive followed by the isolates from *G. zonatum* and *G. boninense* of BSR.

Table 5.8. Materials, equipment and tools needed for inoculation of RWBs

Materials	Equipment and Tools
Ethanol 70%	A spirit lamp
Ganoderma pure culture	Inoculation tool
Sterilized RWBs	Laminar flow cabinet
	Rubber band
	Tissue

Steps in the inoculation of rubber wood blocks

Step 1
Rubber wood block inoculation is normally conducted one day after RWB preparation. The *Ganoderma* pure culture should be ready for use (Section 5.3).

Step 2
Before starting to inoculate the RWBs, all surfaces of relevant equipment and tools must be cleaned with a disinfectant solution or 70% ethanol. Keep the blade of the inoculation tool sterile, flame it with a spirit lamp.

Step 3
Cut the pure isolate into small pieces (5x5 mm) see Fig 5.1.4. Open stopper by removing the aluminium foil/ newspaper that covers the pipe, spray the aluminium foil/newspaper and stopper with 70% ethanol before opening the stopper.

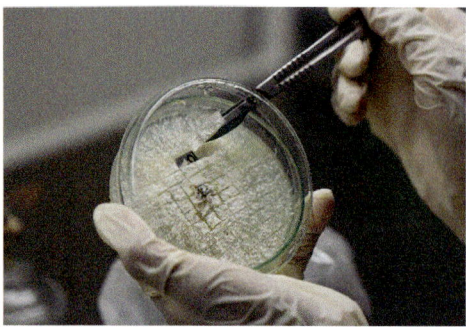

Fig. 5.14. Pieces of *Ganoderma* pure isolates

Step 4
Take 6–8 slices of pure isolate, or for faster results use 1 Petri plate culture of *Ganoderma* pure isolates for 4 bags, place into a plastic bag with a RWB and spread the isolate pieces over all surfaces of the RWB.

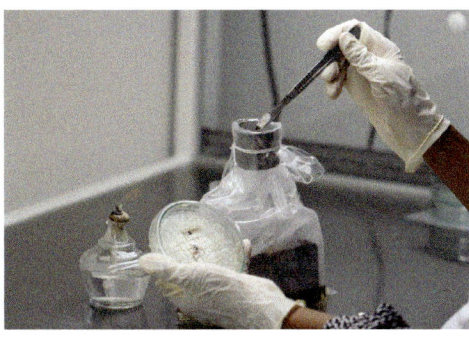

Fig. 5.15. Isolation of *Ganoderma* to RWBs

Step 5
Plug the bags with a stopper.

Step 6
Spray the stopper using 70% ethanol, wrap the bags with aluminium foil and close securely using a rubber band.

Step 7
The RWBs are ready to be incubated and stored in a culture room at 27–30°C in the dark for 12–16 weeks to obtain optimum growth of mycelia.

Step 8
Labelling: write the trial name, isolate name (*Ganoderma* source), and inoculation date on the plastic bag.

5.6   Incubation of Rubber Wood Blocks with *Ganoderma*

Incubation of rubber wood blocks (RWBs) is conducted directly after inoculation. RWBs need to be incubated for 12–16 weeks to build up sufficient inoculum in 6x6x6 cm or 6x6x3 cm RWBs. The time needs to be determined for other RWB sizes. The incubation room should be kept dark except when working in the room. It is recommended that the incubation room is set up with incubation racks able to incubate more RWBs. Iron racks with acrylic glass shelves are recommended as iron racks are strong and durable and acrylic glass is light and has a clear easy to clean surface, so that it may be cleaned and checked easily.

Table 5.9. Materials, equipment and tools needed for incubation of RWBs

Materials	Equipment and Tools
Ethanol 70%	Pen
Inoculated RWBs	Trolley

Steps in the incubation of rubber wood blocks

Step 1
All surfaces of the incubation rack must be cleaned with a disinfectant solution or 70% ethanol.

Fig. 5.16. *Ganoderma* inoculum on rubber wood blocks

Step 2
Place the inoculated RWBs onto a trolley and transfer them to a dark culture room/incubation room at ambient temperature (27-30°C), with air circulation (e.g. ceiling fans).

Step 3
Label: screening trial name, size of RWB, RWB inoculation date, isolate name, and field inoculation date on the rack. Wait for 12-16 weeks (with periodic observations to check what is happening) for the RWBs to be ready for seedling screening (Fig. 5.17).

Fig. 5.17. *Ganoderma* inoculum on rubber wood blocks

Table 5.10. Media and cultures used for routine screening

Materials	Quantity
Rubber wood blocks	3,200 blocks (3,000 + 5% contingency for replacing dead/missing seedlings)
Potato agar (50 cc PA per RWBs bag)	80 L
Pure *Ganoderma* isolates	400 Petri dishes

5.7 Pathogenicity Screening of *Ganoderma* Isolates

In order to obtain the best (most aggressive) *Ganoderma* isolates which can then be used in the screening of oil palm germplasm, it is recommended to screen a range of *Ganoderma* isolates from various sources. *Ganoderma* may be isolated from different palms and different locations. *Ganoderma* isolates on RWBs are screened for aggressiveness in seedling tests; this requires a massive amount of *Ganoderma* inoculum. A *Ganoderma* isolate growing on one RWB is used for one seedling.

The success of nursery screening relies on a set of parameters, one of them is the effectiveness of the inoculum source. Various parameters such as the selection of isolates, material for the substrate, the volume of substrate and the length of incubation time for the fungal isolates being tested, etc. need to be standardized in order to provide robust and reproducible results.

Currently, the use of rubber wood blocks as a substrate for *Ganoderma* is the most effective method of exposing and infecting seedlings with *Ganoderma*. Detailed methods on how to artificially inoculate oil palm seedlings in the nursery with *Ganoderma* are given in Rahmaningsih *et al.*, 2018 (this series); these methods are similar to those of *Trichoderma* screening given in Chapter 7.

Bibliography

Breton, F., Hasan, Y., Hariadi, Lubis, Z. and De Franqueville, H. (2006) Characterization of parameters for the development of an early screening test for basal stem rot tolerance in oil palm progenies. Journal of Oil Palm Research Special Issue April 2006, 24-36.

Idris, A.S., Kushairi, D., Ariffin, D. and Basri, M.W. (2006) Technique for inoculation of oil palm germinated seeds with *Ganoderma*. MPOB Information Series June 2006.

Rahmaningsih, M., Virdiana, I., Bahri, S., Anwar, Y., Forster, B.P. and Breton F. (2018) Nursery *Screening for* Ganoderma *Response in Oil Palm Seedlings: A Manual. Techniques in Plantation Science.* Forster B.P. and Caligari, P.D.S. (eds). CAB International, Wallingford, UK.

Rakib, M.R.M., Bong, C.F.J., Khairulmazmi, A. and Idris, A.S. (2015) Aggressiveness of *Ganoderma boninense* and *G. zonatum* isolated from upper- and basal stem rot of oil palm (*Elaeis guineensis*) in Malaysia. Journal of Oil Palm Research 27(3), 229-240.

In vitro Trichoderma Antagonism Screening

6

Abstract

A protocol is described of how to carry out an *in vitro* test to assess *Trichoderma* antagonism against *Ganoderma*. This is known as a Dual Culture Test. Dual culture is a simple method which, as implied, involves growing *Trichoderma* along side *Ganoderma* in a culture plate. The medium used to grow both the *Trichoderma* and *Ganoderma* is PSA medium (plus antibiotics). This chapter explains the method of how to set up the dual plate test and how to assess *Trichoderma* isolate effectiveness in antagonizing/killing *Ganoderma*.

6.1 Dual Culture Testing

A dual plate test is done by placing a mycelial plug of *G. boninense* against the test isolates (*Trichoderma*) in one culture plate. The antagonistic potential of *Trichoderma* isolates is assessed every day to check the growth of both pathogenic and antagonist fungus, this is done basically by measuring the growth, retardation of growth and death of *Ganoderma*. The method is given below.

Steps in dual culture testing

Step 1
Prepare the PSA plus antibiotics medium (see Section 5.2.1).

Step 2
Grow both *Ganoderma* and Trichoderma in PSA medium (plus antibiotics) in Petri dishes. Petri dish cultures are recommended because they have a large surface area and it is easy for mycelia to be removed by taking out a plug using a cork-borer.

Step 3
Remove mycelial of the *Ganoderma* culture using a cork-borer (0.5 cm), transfer the plug aseptically onto a fresh PSA medium in a Petri dish (9 cm diameter). Then repeat with a *Trichoderma* plug. The plugs should be positioned 3 cm apart as shown in Figure 6.1 and incubated at room temperature (28 ±2°C) with alternate light and dark condition for 8 days (Fig. 6.1).

Fig. 6.1. Dual plate test of *Ganoderma* and *Trichoderma* isolates

Step 4
The radial fungal growth of both fungi are measured every day. The radius of the pathogenic (*Ganoderma*) colony and its retreat from the antagonist (*Trichoderma*) colony are measured (R1).

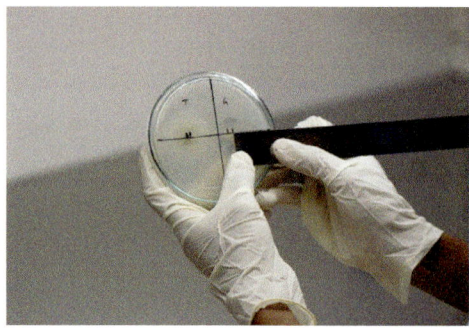

Fig. 6.2. Growth measure of *Ganoderma* mycelia

Step 6
The results are transformed into percentage inhibition of radial growth (PIRG) in relation to radial growth of the pathogen in the plate (R1), using the following formula developed by Skidmore and Dickinson (1976), where

$$PIRG = \frac{R1 - R2}{R1} \times 100$$

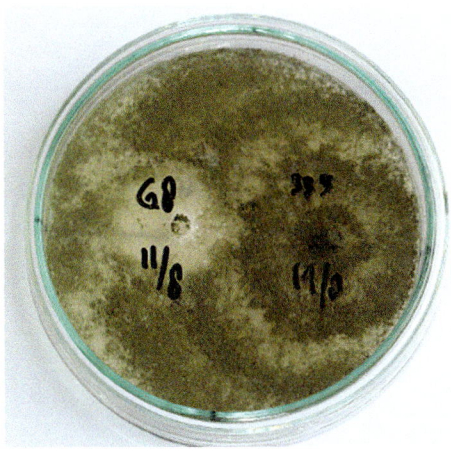

Fig. 6.3. *Trichoderma* inhibiting growth of *Ganoderma* mycelia

Bibliography

Skidmore, A.M. and Dickinson, C. H. (1976) Colony interactions and hyphal inter-ference between *Septoria nodorum* and phylloplane fungi. Transactions of the British Mycolgical Society 66, 161-163.

Sunarwati, D. and Yoza, R. (2010) The ability of Tr ichoderma and Penicillium to inhibit the growth rate of fungus caused Durio root rot (*Phytophthora palmivora*) In vitro. Seminar Nasional Program dan Strategi Pengembangan Buah Nusantara Solok, November 2010, 176-189.

Sundram, S. (2014) First report: Isolation of endophytic *Trichoderma* from oil palm (*Elaeis guineensis* Jacq.) and their *in vitro* antagonism. Journal of Oil Palm Research 25(3), 368.

Zivkovic, S., Stojanovic, S., Ivanovic, Z., Gavrilovic, V., Popovic, T. and Balaz, J. (2010) Screening of antagonistic activity of microorganism against *Colletotrichum acutatum* and *Colletotrichum gloeosporioides*. Archives of Biological Sciences 62(3), 611-623.

Trichoderma Nursery Screening for *Ganoderma* Control

7

Abstract

Nursery screening of seedlings is currently the fastest and easiest method to screen the realistic biopesticide effectiveness. The nursery screening process takes about 6 months to 1 year. Certain critical parameters are required for efficient nursery screening, including: shading, *Ganoderma* rubber wood block quality, inoculation stage, the inoculation process and environment conditions. As with laboratory procedures, each parameter in the nursery should be standardized as much as possible as this minimizes the variation between and within trials. Step by step methods in the various stages of nursery screening are described: experimental design, nursery preparation, selection of *Ganoderma* inoculated rubber wood blocks (RWBs), preparation of *Trichoderma* and seed/seedling inoculation.

7.1 Nursery Preparation

The nursery may be a temporary or permanent structure, typically supported by iron or bamboo poles. This structure supports shading, which is the most important parameter for *Ganoderma* screening of oil palm seedlings. Experiments demonstrated that 90% shade gives the highest infection to the inoculated oil palm seedlings as *Ganoderma* grows rapidly under dark conditions (Rahmaningsih *et al.*, 2018). Ambient temperatures can also affect the success of seedling inoculation. Although *Ganoderma* development is dependent on soil temperature, ambient air temperature in the nursery also has a significant impact on the temperature inside the polybag in which the seedling is grown. The maximum recommended temperature inside the shade nursery is 30°C.

Fig. 7.1. *Ganoderma* seedling screening nursery with shading: left inside view, right outside view

Table 7.1. Materials, equipment and tools needed for nursery preparation

Materials	Equipment and Tools
Sieved soil (top soil) Water for watering	Nursery construction with net shading Paint Polybags size 20 x 30 x 0.1 cm Soil sieve

Steps in preparing the nursery

Step 1

Before starting a screening trial, preparation of trial design with respect to purpose and statistical rigour is required. This is normally a replicated randomized block design. Treatments include *Ganoderma* (inoculated RWBs) and *Trichoderma* (inoculated medium) while controls have no *Trichoderma*, but with planted germinated seeds and *Ganoderma* infected RWBs. It is preferable to use seeds and seedlings from a single progeny (or similar progeny, or better, a randomized representative sample of progenies – such as commercial seedlings) to standardize host genetic variation for *Ganoderma* susceptibility.

An example layout is given below. This is an example of a trial consisting of five treatments: A, B, C, D and E (four of *Trichoderma* isolates and one control) with three replicates. One treatment can consist of 16, 20, 30 and more seedlings.

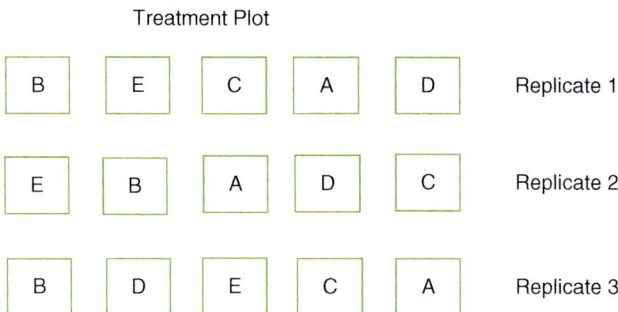

Fig. 7.2. An example of trial layout.

Step 2
About two weeks before seed/seedling planting, polybags need to be partially filled with soil. Fresh top soil from a non-*Ganoderma* area should be used. Soil should be sieved to remove root and other debris. Soil is placed into the polybags to a depth of about 10 cm. There should be an empty volume above the soil to place the rubber wood block (RWB) inoculated with *Ganoderma* and the soil (top 14 cm of the polybag, see Fig. 7.3).

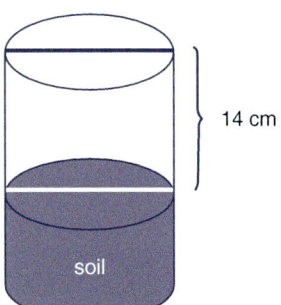

Fig. 7.3. Preparation of polybags, initial (partial) soil fill

Step 3
The prepared polybags are then arranged in plots according to the randomization of the experimental design.

Step 4
Label the polybags with the treatment and number of polybags based on trial map. Place labels on polybags to identify replicate and plot (Fig. 7.4).

Fig. 7.4. Labelled polybag with treatment and plot number

7.2   Selection of Inoculated Rubber Wood Blocks

Rubber wood blocks (RWBs) are usually incubated in the dark room (Chapter 5, see also Rahmaningsih *et al.*, 2018). The selection/rejection of inoculated RWBs is done weekly, rejection is based on the existence of contaminants (bacteria and other fungi) and limited growth of *Ganoderma* on the RWBs. Inoculated RWBs with heavy or light contamination should also be rejected. In addition, RWBs with less than 90% *Ganoderma* mycelia covered are rejected (Fig. 7.5). Training should be given to the workers in selecting RWBs for seedling screening.

Fig. 7.5. Variation in *Ganoderma* coverage on RWBs from right to left high coverage (100%); medium coverage (50%) and poor coverage (<10%)

Table 7.2. Materials, equipment and tools needed in selecting inoculated RWBs

Materials	Equipment and Tools
Inoculated RWBs	Torch for inspection in the dark
Plastic bag to keep the contaminated or un-growth isolate	

Steps in selecting inoculated RWBs

Step 1

Two weeks after inoculating RWBs with *Ganoderma*, inspection can begin to check the growth of the isolates and the presence of other fungal or bacterial contamination. These checks should be done weekly.

Step 2

The inoculum source RWBs inoculated with *Ganoderma* should be selected one week before the test germinated seeds/seedlings are planted. The selected RWBs should have over 90% mycelial growth.

Step 3

Before inoculated RWBs are placed into polybags, the RWBs should be rechecked for contamination by bacteria or other fungi. Only RWBs with >90% mycelia growth is used.

Fig. 7.6. Selected RWBs with >90% *Ganoderma* coverage and no contamination

7.3   Preparation of *Trichoderma* Isolates

Stocks of *Trichoderma* isolates (see Chapters 3 and 4) to be used for testing should be weighed according to the treatment dosage. The dosage (see below) used is normally 100–200 g per polybag for nursery screening. Labelling should be done for each isolate.

Table 7.3. Materials, equipment and tools needed for preparation of *Trichoderma* isolates

Materials	Equipment and Tools
Trichoderma isolates	Gloves
	Safety mask
	Scale
	Marker
	Plastic bag

Step 1
Two weeks before application, ensure all *Trichoderma* media for testing are ready. See Chapter 4. Check the expiry date for each of the isolates.

Fig. 7.7. *Trichoderma* media ready for trialling

Step 2
Each isolate is then weighed according to the treatment dosage. Usually 200 g per polybag is used. This can be mixed with soil (See Step 3 of Nursery Inoculation).

Fig. 7.8. Weighing out *Trichoderma* medium for soil inoculation

Step 3

Place each weighed isolate in a plastic bag and label with the isolate and treatment codes.

Fig. 7.9. *Trichoderma* is labelled and ready for soil inoculation

7.4   Nursery Inoculation

RWBs are taken from the incubation room at the laboratory and transferred to the nursery, to be placed into the prepared polybag. Polybag placement (experimental design) can be done one day to one week prior to planting. The distance between the surface of RWB and the surface of polybag (normally 14 cm) is critical and should be standardized and checked carefully.

Table 7.4. Materials, equipment and tools needed for nursery inoculation

Materials	Equipment and Tools
Polybag 20x30x0.1 cm with soil	Calibration tool
Selected inoculated RWBs	Gloves and face mask when handling *Trichoderma*

Step 1
Open the package containing the inoculated RWBs.

Step 2
The selected *Ganoderma* inoculated RWB is then placed onto the soil in the prepared polybag at 14 cm from the top of the bag so that the upper surface of the block is about 8 cm from the top of the polybag (Fig 7.10). A calibration tool should be used to standardize the distance between the RWB's top surface and the top of the polybag (Fig. 7.10). Then cover the inoculum source with top soil (Fig. 7.11) to a depth of 8 cm.

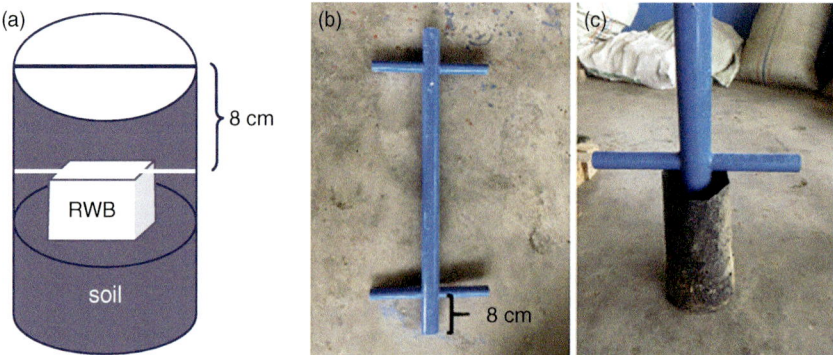

Fig. 7.10. (a) Distance between RWB surface and polybag surface. (b) Calibration tool. (c) Distance standardization using a calibration tool

Step 3

Prepare *Trichoderma* treatments. Medium containing *Trichoderma* (see Chapter 4) can be applied by mixing with soil. *Trichoderma* is mixed with soil by hand (wear gloves and mask) then add mixture to polybags above RWB to a depth of 8 cm (Fig. 7.10 above) before seed/seedling planting (see below).

Fig. 7.11. Mixing *Trichoderma* with soil

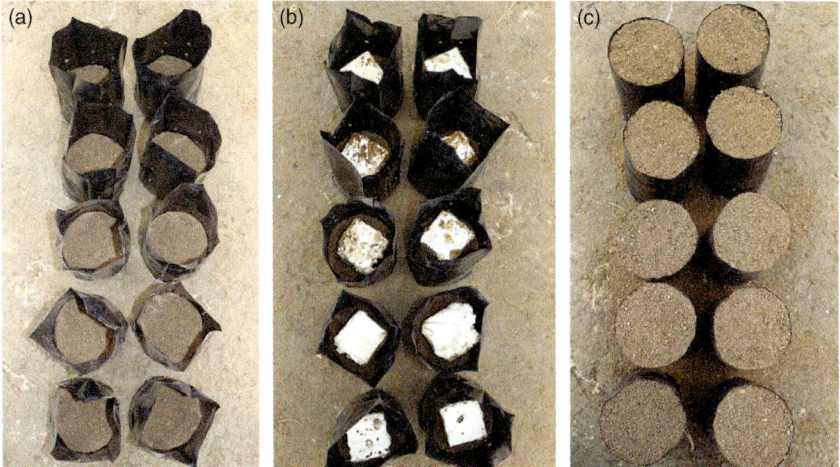

Fig. 7.12. Stages in polybag preparation prior to seed/seedlings planting: (a) Polybags filled with soil to a depth of 10 cm. (b) RWB placement in the middle of polybag. (c) Polybag topped up with *Trichoderma* inoculated soil

7.5 Planting

Planting is an important stage in *Trichoderma* nursery screening. Materials that are used for nursery screening should be either germinated seeds or young (three-month-old) seedlings. In order to minimize root injury during this stage, the use of germinated seeds is more favourable, and easier to handle. In addition, disease symptoms can be seen more rapidly using germinated seeds.

Uniform germinated seeds or seedlings need to be prepared (see Kelanaputra *et al.*, 2018) and ready for planting. In order to standardize plant response to *Ganoderma* it is preferable that seedlings come from random sample of seeds or commercial seedlings.

Table 7.5. Materials, equipment and tools needed for planting

Materials	Equipment and Tools
Oil palm germinated seeds	Pen
Oil palm seedlings	Trial Board
	Watering can

Step 1
One or two days before planting, re-check the polybags and plot arrangement. Spare polybags are also needed to be set up as reserve in case the seedling fails to emerge.

Step 2
One day before planting, polybags are watered using a watering can for small scale of trials (approx. 200–300 ml per polybag).

Fig. 7.13. Hand watering prepared polybags

Step 3
On the day of planting, all planting materials should be at the nursery along with identification information such as bunch reference.

Step 4
Record the name of treatment/progeny code of the plant material. Check that the information on the label attached to the seed bag matches that of the received seedlings/germinated seed.

Step 5
Plant oil palm seedlings (3 months old) or germinated seeds into polybags based on trial design, one seed/seedling per polybag. This is done manually and carefully by pushing the seed through the soil to the RWB, so as not to damage the seeds. For three month old seedlings, these are removed gently from their polybags and loose soil removed by shaking. The roots are then gently twisted together and pushed through the soil and placed on top of the RWB. The control treatments should be set up first followed by *Trichoderma* treatment plots.

Fig. 7.14. Germinated seeds ready to be planted on polybags

Step 6
Record information of trial number, planting date, name of isolate, progeny code (if any) on a plate in front of the trial plots.

Step 7
Record information on *Ganoderma* infection (Rahmaningsih *et al.*, 2018) every two weeks or monthly.

Bibliography

Anjara, G., Virdiana, I., Flood, J., Ritchie, B.J. and Nelson, S. (2011) Preliminary in vitro and nursery results to screen for *Trichoderma* isolates antagonistic to *Ganoderma*. In: Proceedings of the International Palm Oil (PIPOC) Congress, Agriculture, Biotechnology and Sustainability. 15th-18th November. Kuala Lumpur. Published by Malaysian Palm Oil Board. P 424-427.

Breton, F., Hasan, Y., Hariadi, Lubis, Z. and De Franqueville, H. (2006) Characterization of parameters for the development of an early screening test for basal stem rot tolerance in oil palm progenies. Journal of Oil Palm Research Special Issue April, 24-36.

Kelanaputra, E.S., Nelson, S.P.C., Setiawati, U., Sitepu, U., Nur, F., Forster, B.P. and Purba, A.R. (2018) *Seed Production in Oil Palm: A Manual. Techniques in Plantation Science.* Forster B.P. and Caligari, P.D.S. (eds). CAB International, Wallingford, UK.

Rahmaningsih, M., Setiawati, U., Breton, F., Sore, R., Nelson, S. and Caligari, P.D.S. (2010) General combining ability (GCA) estimates for *Ganoderma* partial resistance from nursery screening trials. In: Proceedings of 2010 International Oil Palm Conference, 1-3 June 2010, Yogyakarta, Indonesia.

Rahmaningsih, M., Virdiana, I., Bahri, S., Anwar, Y., Forster, B.P. and Breton F. (2018) Nursery *Screening for* Ganoderma *Response in Oil Palm Seedlings: A Manual. Techniques in Plantation Science.* Forster B.P. and Caligari, P.D.S. (eds). CAB International, Wallingford, UK.

Virdiana, I., Anjara, P., Flood, J., Sitepu, B., Hasan, Y., Aditya and R., Nelson, S. (2012) Replanting system and *Trichoderma* as an effective form of *Ganoderma* control and proven results. 4th Oil Palm Summit. 09-10 July. Denpasar. P1-10.

Scoring Response of *Ganoderma* to *Trichoderma*   **8**

Abstract

Standard methods in screening seedling response to *Ganoderma* have been described by Rahmaningsih *et al.*, 2018 (this series). Here we describe methods to score the effectiveness of *Trichoderma* isolates as antagonists of *Ganoderma* through observations on seedlings infected by *Ganoderma*. This is done in the nursery by growing seedlings in soil inoculated with *Ganoderma* RWBs and *Trichoderma* (see Chapter 7). Two different observations of seedlings are conducted, one is external, the other is internal. For the external observations, the appearance of *Ganoderma* mycelium, basidiocarp and/or foliar symptoms (including seedling mortality) are recorded. For internal observations, seedlings are considered infected by *Ganoderma* when decayed (rotten) tissues are observed inside the bole. External observations are conducted every two weeks or monthly while the internal symptoms are observed at the end of each experiment. The first observations are made during the first month, and thereafter once a month until the first *Ganoderma* symptoms are detected, thereafter observations are carried out more regularly, every two weeks.

8.1 External Observations

External observations follow the methods described by Breton *et al.* (2006) and Idris *et al.* (2006). Foliar discoloration (leaf browning, leaf drying from older to younger leaves and finally leaf death) is counted as an infection. Usually *Ganoderma* foliar symptoms are followed by the appearance of mycelium and/or basidiocarps (fruiting bodies) on seedlings or on the side of polybags. Standardization of the various symptom scores are important in an effort to develop rigour to the screening. Personnel who conduct the observations should understand the disease symptoms in seedlings and how to score them.

Table 8.1. Materials, equipment and tools needed for external observation

Materials	Equipment and Tools
Oil palm seedlings inoculated with *Trichoderma* and *Ganoderma*	Observation form (Fig. 8. 1) Clip board Pen

PT TIMBANG DELI INDONESIA
VERDANT BIOSCIENCE

Verdant BIOSCIENCE

DISEASE RECORD DATA / GANODERMA ON NURSERY RECORD DATA

Expt No:
Date:

Recorder:
Estate/Div/Block:

Plot	Palm No	Rep	Infected by Ganoderma	Died by Ganoderma	FB	M	LS	MS	FBS	FBP	Plot	Palm No	Rep	Infected by Ganoderma	Died by Ganoderma	FB	M	LS	MS	FBS	FBP
1											1										
2											2										
3											3										
4											4										
5											5										
6											6										
7											7										
8											8										
9											9										
10											10										
11											11										
12											12										
13											13										
14											14										
15											15										
16											16										

Basidiocarp Appearance (Yes/No) spans the FB, M, LS, MS, FBS, FBP columns.

Fig. 8.1. An example of an observation form for recording *Ganoderma* disease symptoms in seedlings

Steps in taking external observations

Step 1
The entire experimental plot is inspected regularly (being mindful of weather conditions, wind and rain that may disturb the polybag layout). Each polybag is observed regularly for seedling emergence.

Step 2
Seedlings are expected to emerge from the soil after 10 to 14 days after planting.

Step 3
First observations are normally carried out at one month after planting when all seedlings should have emerged. Seedlings which have failed to emerge, and any abnormal seedlings are identified; these seedlings should be replaced by the normal seedlings from the spare plot (see Chapter 7) and the fact

recorded. External observations are conducted every two weeks or monthly while the internal symptoms are observed at the end of each experiment.

Step 4
Seedlings are identified based on the information given on their label attached to the polybag.

Fig. 8.2. Routine external observations taken in the nursery

Step 5
All symptoms are recorded on the observation form (Fig. 8.1).

Fig. 8.3. External observation symptoms consist of healthy seedling, infected seedling with *Ganoderma* mycelia and infected seedling with basidiocarp (from left to right)

Step 6
External observations are completed usually about 8 to 10 months after planting, or when the control plot has >80% infection.

8.2   Internal Observations

Internal observations are conducted in the nursery using destructive tests on the seedlings. Seedlings are cut longitudinally using a knife. The severity of internal symptoms is assessed by a visual estimation of amount of damaged tissue caused by *Ganoderma* using a scale established by Breton *et al.* (2006).

Fig. 8.4. Longitudinal section of a seedling showing internal *Ganoderma* rot symptoms at the base of the bole

Table 8.2. Materials, equipment and tools needed for internal observation

Materials	Equipment and Tools
Oil palm seedlings inoculated with *Ganoderma*	Knife Observation form (Fig. 8.1) Pen

Steps in taking internal observations

Step 1
At the end of the screening period (about 8 to 10 months, or when control plot (no *Trichoderma* added) has >80% infection), each seedling should be

removed from its polybag and cut open longitudinally with a knife, splitting the bole into equal halves.

Step 2
Identify both healthy and infected seedlings. Infected seedlings have discolouration and decay of the bole tissues.

Step 3
All the symptoms are recorded on the observation form (Fig. 8.1).

Step 4
The experiment is concluded and materials disposed carefully. Seedlings, soil and RWBs are removed from the polybags. Polybags are removed and disposed of in a special plastic dump, soil is placed in a confined area. The RWBs and seedlings should be incinerated.

Outcomes

Nursery screening is a good method to test the best *Trichoderma* isolates. This is simpler and requires shorter time than field testing. The tests can be used to discover more potent sources of *Trichoderma* for commercial production, i.e. testing against standard commercial isolates to find better (more aggressive) isolates of *Trichoderma*, which may then be developed for commercial production (see Chapter 7).

Bibliography

Breton, F., Hasan, Y., Hariadi, Lubis, Z. and de Franqueville, H. (2006) Characterization of parameters for the development of an early screening test for basal stem rot tolerance in oil palm progenies. Journal of Oil Palm Research Special Issue April 2006, 24-36.

Idris, A.S., Kushairi, D., Ariffin, D. and Basri, M.W. (2006) Technique for Inoculation of Oil Palm Germinated Seeds with *Ganoderma*. MPOB Information Series. ISSN 1511-7871 2006, 1-3.

Rahmaningsih, M., Virdiana, I., Bahri, S., Anwar, Y., Forster, B.P. and Breton F. (2018) Nursery *Screening for* Ganoderma *Response in Oil Palm Seedlings: A Manual. Techniques in Plantation Science.* Forster B.P. and Caligari, P.D.S. (eds). CAB International, Wallingford, UK.

Production of *Trichoderma* for Commercial Application

9

Abstract

From nursery screening (Chapter 8) the most effective *Trichoderma* isolates can be identified and advanced for large scale, commercial production and application. In order to use *Trichoderma* on a large or commercial scale, huge stocks need to be produced and stored ready for field application. Stock have a limited storage time and are assigned an expiry date, for Verdant this is eight months. Stocks of *Trichoderma* are normally produced six months before soil application. *Trichoderma* is most often used in re-planting, i.e. continuous oil palm production as *Ganoderma* incidence is high. *Trichoderma* is stored in a go-down with a concrete floor and good aeration and with road access for transportation to the field when required.

Table 9.1. Materials, equipment and tools needed for commercial application

Materials	Equipment and Tools
Rice husk media	
Trichoderma in maize and rice husk medium	Pail
	Labelled gunny bag
	Weighing scale
	Room or Go-down
	Burlap sewing machine
	Burlap sewing thread
	Permanent marker
	Book
	Pen
	Transport by truck

Steps in commercial *Trichoderma* production

Step 1
Large scale *Trichoderma* cultures in maize medium are set up for large scale mycelia biomass production. Details of *Trichoderma* multiplication are given in Chapter 4.2. *Trichoderma* Multiplication on Maize Medium.

Step 2
Mix *Trichoderma* in maize medium with rice husk medium in a ratio of 1:4 as follows. Prepare 2 kg of *Trichoderma* in maize medium and weigh 8 kg of rice husk medium. Place both rice husk and *Trichoderma* in a pail (diameter 50 cm) and mix by hand.

Step 3
Trichoderma in bran medium is poured directly from the pail to gunny bag, 30 kg per gunny bag.

Step 4
The gunny bag should be closed securely to avoid insect infestation, this is done by closing the top of the gunny bag and sealing it by sewing using a burlap sewing machine with burlap sewing thread.

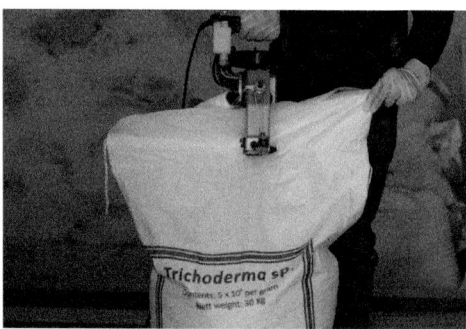

Fig. 9.1. Manual sewing of *Trichoderma* gunny bag

Step 5
A printed label is attached to the gunny bag (using an inside and outside plastic layer). The label should have information on the product name, contents, weight, production date and expiry date. The production date and expired date is write manually using a permanent marker pen.

Step 6

Place *Trichoderma* gunny bags on wooden pallets on a concrete floor inside a clean room with aeration. The room needs to be free of insects and should be a designated *Trichoderma* storage room – there should be no chemicals in the storage area.

Step 7

Document the *Trichoderma* stock in a record book which is maintained for each *Trichoderma* stock with details of date and amount of product sent to the field for application.

Fig. 9.2. *Trichoderma* product in both closed and open bag

Step 8

Trichoderma is usually delivered to the field using a truck. For application in the field see Chapter 10.

Bibliography

Panahian, G.H., Rahnama, K. and Jafari, M. (2012) Mass production of *Trichoderma* spp. and application. International Research Journal of Applied and Basic Sciences 3(2), 292-298.

Kumar, S., Thakur, M. and Rani, A. (2014) *Trichoderma*: Mass production, formulation, quality control, delivery and its scope in commercialization in India for the management of plant disease. African Journal of Agricultural Research 9(53), 3838-3852.

Trichoderma Application in Oil Palm Plantations

10

Abstract

Trichoderma applications to help protect oil palm from *Ganoderma* can be carried out in the field. This is done during the field planting process. Here we describe the procedures involved, from the preparation of the *Trichoderma* product and how it is applied to the planting hole prior to planting young palm plants.

10.1 *Trichoderma* Application in the Field

Field application of *Trichoderma* is done at the time of planting young oil palm plants (ex seedlings or ramets), which are prepared in the nursery (see Laksono *et al.*, 2019, this series). The preparation of the planting area is described by Sitepu *et al.*, 2019 (this series). Here we describe activities once the planting holes have been made and the planting materials have arrived from the nursery ready for planting. Three different activities are done in series: 1) application of slow release Rock Phosphate (RP) fertilizer to the planting hole, 2) application of *Trichoderma* product (see Chapter 9) to the planting hole and top soil, and 3) planting the young oil palm into the planting hole. The hole should be at least 40 cm depth and dug using a hole-digger as per estate standard practice (see Sitepu *et al.*, 2019).

Table 10.1. Materials, equipment and tools needed for field application

Materials	Equipment and Tools
Trichoderma product	Hole-digger
Field ready young palms (seedlings or ramets)	Hoe
	Bags for pre-weighed *Trichoderma* product per planting hole, or measuring cup
Soil	
Rock phosphate	

Steps in *Trichoderma* application in the field

Step 1

Prepare area to be planted with oil palm and dig planting holes (see Sitepu *et al.*, 2019 for details). *Trichoderma* product can be applied as soon as each hole is dug.

Fig. 10.1. Oil palm planting hole

Step 2

The *Trichoderma* product is pre-weighed, e.g. 200 g per planting hole and placed into separate bags (one per planting hole); alternatively a measuring cup for 200 g *Trichoderma* product can be used.

Step 3

Apply 800 g Rock Phosphate (RP) fertilizer per hole by hand. The RP is normally pre-weighed into bags (one bag per planting hole). Half of the RP (400 g) is applied to the inside of the planting hole, and the balance (400g) is mixed with top soil used to cover the planting hole after the young oil palm plant has been placed into the hole.

Fig. 10.2. Rock phosphate application into planting hole

Step 4

After applying 400 RP around the inside of the planting hole, add a layer of about 5 cm of top soil mixed with RP to the base of the hole. Then apply 200 g *Trichoderma* product inside the planting hole and around the wall; this is done by hand.

Fig. 10.3. *Trichoderma* applied in the planting hole

Step 5

Remove the polybag from the young plant taking care not to disturb the roots. Place the young oil palm plant into the hole (see Sitepu *et al.*, 2019 for details and labour requirements).

Fig. 10.4. Planting a young oil palm plant into the prepared planting hole

Step 6

Fill the hole with the top soil/ RP mix until it is half full, and then carefully gently compact the soil all-round the young oil palm plant by stamping.

Step 7

Continue to fill the hole until it is full and then once again gently compact the soil all-round the palm.

Fig. 10.5. Compact the soil all-round the seeding

Step 8
Remove the string from the palm.

Step 9
Planting is recommended to be done in the rainy season. If the planting is done in the dry season, then watering is required.

10.2 Post-application Monitoring

The planting work is normally supervized and checked to ensure all work has been done correctly and in accordance with SOPs.

Surveys for pests and diseases normally start one month after planting. Any missing or diseased plants are noted and should be replaced by 'supply' plants from the nursery (see Laksono *et al.*, 2019). *Ganoderma* censuses are done twice a year.

10.3   *Trichoderma* and *Ganoderma* Monitoring in the Soil

There have been a number of genomic studies to investigate the interaction between oil palm, *Ganoderma* and *Trichoderma*. However, a major limitation of *Ganoderma* detection is that the physical symptoms do not appear until the disease is at a critical stage and these symptoms can take a few years to develop. Genomics tools are being developed to detect and monitor *Ganoderma* in both soil and in palms. DNA techniques are also being used to genotype *Trichoderma* and to assess the potential of different genotypes as a biocontrol. Several detection methods have been used in attempts to detect *Ganoderma* infection as early as possible, ranging from immunoassay-based (the first molecular attempts to detect *Ganoderma*) to DNA-based testing (Hushiarian *et al.*, 2013).

Genomic identification of fungi is usually done using DNA sequences within the Internal Transcribed Spacer 1 and 2 (ITS1 and 2) region (Schoch *et al.*, 2012). The ITS region is a universal marker for fungi, thus identifications based on ITS sequence data alone must be used with caution, especially when dealing with closely related species. An alternative is to use sequence of the protein-coding gene translation-elongation factor 1-α (tef) for *Trichoderma* and *Ganoderma* identification, since this is more variable than ITS rDNA (Samuels, 2005; Loyd *et al.*, 2018).

In addition to *Ganoderma* and *Trichoderma*, it is known that there are several groups of microorganisms which affect soil quality in oil palm plantations, thus soil monitoring is needed to assess the soil health. Soil monitoring is done before planting/replanting as a baseline point and then repeated periodically during the life of the plantation. Soil DNA analyses can compare the amount of *Ganoderma*, *Trichoderma*, and other beneficial and pathogenic microorganisms within the soil over time and thus provide growers with vital information on soil health (Kirk *et al.*, 2004). Verdant is currently developing DNA diagnostics to monitor soil health and a manual on soil DNA techniques is in preparation for this series.

Bibliography

Hushiarian, R., Yusof, N.A., Dutse, S.W. (2013) Detection and control of *Ganoderma boninense*: Strategies and perspectives. *SpringerPlus*, 555 (2), 1-12.

Jambhulkar, P.P., Sharma, P., Meghwal, M.R. (2015) Additive effect of soil application with *Trichoderma* enriched FYM along with seed treatment and drenching with *Trichoderma* formulation for management of wet root rot caused by *Rhizoctonia solani* in chickpea. Journal of Pure Applied Microbiology 9(1), 8 pp.

Kirk, J.L., Beaudette, L.A., Hart, M., Moutoglis, P., Klironomos, J.N., Lee, H., Trevors, J.T. (2004) Methods of studying soil microbial diversity. *Journal of Microbiological Methods*, 58, 169-188.

Laksono, N.D., Setiawati, U., Nur. F., Rahmaningsih, M, Anwar, Y. *et al.* (2019) *Nursery Practices in Oil Palm: A Manual. Techniques in Plantation Science.* Forster BP and Caligari PDS (eds). CAB International, Wallingford, UK, in press.

Loyd, A.L., Barnes, C.W., Held, B.W., Schink, M.J., Smith, M.E., Smith, J.A., Blanchette, R.A. (2018) Elucidating "lucidum": Distinguishing the diverse laccate Ganoderma species of the United States. *PLoS ONE*, 13 (7), 1-16.

Samuels G.J. (2005) Trichoderma: systematics, the sexual state, and ecology. *Phytopathology*, 96, 195-206.

Singh, H.B., Srivastava, S., Singh, A., Katiyar, R.S. (2007). Field efficacy of *Trichoderma harzianum* application on wilt disease of cumin caused by *Fusarium oxysporum* f. sp. cumini. *Journal of Biological Control* 21(2), 317-319.

Sitepu, B., Setiawati, U., Nur, F., Laksono, N.D., Anwar, Y. *et al.* (2019) *Field Trials in Oil Palm breeding: A Manual. Techniques in Plantation Science.* Forster B.P. and Caligari, P.D.S. (eds). CAB International, Wallingford, UK, in press.

Schoch, C.L., Seifert, K.A., Huhndorf, S., Robert, V., Spouge, J.L, Levesque, C.A., Chen, W, Fungal Barcoding Consortium. (2012) Nuclear ribosomal internal transcribed spacer (ITS) region as a universal DNA barcode marker for fungi. *PNAS*, 109 (16), 6241-6246.

Index

Page numbers in **bold** type refer to figures and tables.

CABI – who we are and what we do

This book is published by **CABI**, an International not-for-profit organisation that improves people's lives worldwide by providing information and applying scientific expertise to solve problems in agriculture and the environment.

CABI is also a global publisher producing key scientific publications, including world renowned databases, as well as compendia, books, ebooks and full text electronic resources. We publish content in a wide range of subject areas including: agriculture and crop science / animal and veterinary sciences / ecology and conservation / environmental science / horticulture and plant sciences / human health, food science and nutrition / international development / leisure and tourism.

The profits from CABI's publishing activities enable us to work with farming communities around the world, supporting them as they battle with poor soil, invasive species and pests and diseases, to improve their livelihoods and help provide food for an ever growing population.

CABI is an international intergovernmental organisation, and we gratefully acknowledge the core financial support from our member countries (and lead agencies) including:

Discover more

To read more about CABI's work, please visit: **www.cabi.org**

Browse our books at: **www.cabi.org/bookshop**,
or explore our online products at: **www.cabi.org/publishing-products**

Interested in writing for CABI? Find our author guidelines here:
www.cabi.org/publishing-products/information-for-authors/